Moreton Morrell Site

## Plant *Names Simplified*

A. T. J

and

H. A. SMITH

First published, 1931
Second edition, 1946
Reprinted 2008 from the Landsmans edition, by agreement

ISBN 978-1-905523-82-5

Published by
Old Pond Publishing Ltd
Dencora Business Centre
36 White House Road
Ipswich
IP1 5LT
United Kingdom

www.oldpond.com

Illustration: Liz Whatling
Printed and bound in Great Britain by The Nuffield Press, Abingdon

## PREFACE

MY ENDEAVOUR in preparing this glossary has been to offer the reader a simple translation and pronunciation of the names of such plants, trees and shrubs as are commonly grown in the average garden. As they stand, such names are to most of us something more than an awkward obstacle barring the way to any real intimacy with the elements of botany. They are a direct hindrance to our progress as gardeners, and tend to complicate rather than to elucidate the difficulties of a vast and absorbing subject.

Centuries ago each plant was known by a long, descriptive sentence, which was unwieldy, to say the least. Then Caspar Bauhin (1560–1624) devised a plan of adopting two names only for each plant. But it was not until the great Swedish naturalist, Linnæus (1707–1778), undertook the task of methodically naming and classifying the whole living world "from buffaloes to buttercups" that the dual name system became permanently established. Linnæus brought order out of chaos and indexed the vegetable world on a basis so sound and universally acceptable to the peoples of all nations that most of his names are in use to this day.

Of the two names given to each plant, the first, which may be likened to our surname, is the generic, or group, name. This can occur only once—that is, as a group name—but while the second, the specific name or species is only given to one plant of the same genus—as is a Christian name in a family—it may occur in many different genera. Plants of garden origin, or those which are mere variations from the true species, usually have a third name, such as campanula (generic), rotundifolia (specific), album (varietal). But since most of these varietal names occur as specific names they have not been generally included in the glossary. Names of plants which are hybrids, or which otherwise have had a garden origin, have also been omitted. A very large subject has demanded as much brevity as possible, the writer realising that a dictionary of this kind might defeat its own object were it too bulky. Regarding such specific names as must constantly recur—japonica, chinensis, Wilsoni, vulgare, floribunda, and the like—I have avoided the usual method of a single mention followed by cross-references which is apt to waste the reader's time and patience. Instead of this, such words have been repeated throughout the work in all genera with which they are most closely associated.

Generic names, being the more important, are accorded the fuller explanation, and most of the more familiar groups are afforded rather more liberal treatment than those not so well known. Specific names are treated on similar lines. These are mainly adjectival or descriptive of the plant's colour, form, or habit, but space only allows us to give to each of these a literal translation of the word with a brief indication as to what it alludes. Where a specific name is plainly derived from the name of a person (often merely commemorative) or place, its meaning is usually sufficiently obvious without explanation.

It is felt that with a translation of these words before him, the average amateur will be afforded not only an interesting glimpse into the past history of his plants, but that the names will lose much of that awkwardness with which they are so often invested. Further, the second or specific name, alluding as it so often does to some marked character in the plant, will, in its translated form, often be helpful in aiding the identification of doubtful species. And, after all, these Greek and Latin names which are so disturbing to some people need not in themselves be any more foreign and unintelligible than many other words now recognised as English and in daily use. It is entirely a matter of custom, and when we realise how freely we now use such beautiful botanical names as campanula, veronica, and clematis now that they have become familiar to us, one may with confidence express the belief that the time will come when the multitude of plant names which are not so well known will be spoken with the ease of the examples given. And it is to the furtherance of that desirable end that this glossary has been prepared. Fuchsia, geranium, pelargonium, dahlia, aster, primula, calendula, chrysanthemum, pyrethrum are a few additional botanical plant names that have become absorbed into everyday language.

Regarding pronunciation, this is given by phonetic spelling, each syllable being sounded as it is spelt. But in this matter the writer claims no infallibility. Just as in English (as departures in broadcasting have pointed out), where classical accuracy is sometimes superseded by custom, so is the pronunciation of these Latin and Greek names often decided by common usage. The correct pronunciation of anemone, for example, is an-e-*mo-ne*, with the accent on each of the last two vowels. But we shall continue to say an-*em*-on-e. Hypericum will always be hy-*per*-ik-um, rather than hy-per-*i*-kum; and erica will be *er*-ik-a and not er-*i*-ka, as the learned would have such names said. The fact is no arbitrary rules of pronunciation can be laid down. Even such a great authority as the

*Oxford English Dictionary* will give a choice of two pronunciations for the same word; and when we are given full licence by such exalted powers to pronounce chauffeur as "shofer", fauteuil as "fotill", cinema as "sinnema", and Celtic as "seltik", one may grieve for the traditions of the classics, but give courage its due and adopt in the pronunciation of our botanical names an equally liberal attitude.

Apropos the subject of pronunciation we may quote some lines on "cyclamen" which appeared in the pages of an old-time gardening periodical:

How shall we sound its mystic name
Of Greek descent and Persian fame?
Shall "y" be long and "a" be short,
Or will the "y" and "a" retort?
Shall "y" be lightly rippled o'er,
Or should we emphasise it more?
Alas! The doctors disagree,
For "y's" a doubtful quantity.
Some people use it now and then,
As if 'twere written "Sickly-men";
But as it comes from *kuklos*, Greek,
Why not "kick-laymen", so to speak?
The gardener, with his ready wit,
Upon another mode has hit;
He's terse and brief—long names dislikes,
And so he renders it as "Sykes".

At the same time one must have a working principle as a basis in the pronunciation of scientific plant names, and the most important point, perhaps, is to get the correct vowel sound, or stress, in the right place, and this in each case is indicated by the syllable being printed in italics. With few exceptions the accented vowel (in italics) is long when alone or following a consonant (*o* as in "mole", *a* as in "pate", *e* as in "been"), short when preceding a consonant or between two (*o* as in "on" or "pon", *a* as in "an" or "pan", *e* as in "en" or "pen"). The *u* is given the letter (*ew*) sound, as in "due". The *g* is hard (as in "get"), but soft when the phonetic spelling gives it the *j* sound (as in "gem"), and *c* is hard (as in "can"), when it is given the *k* sound, or soft (as in "pace"), when it is given the sound of *s*.

Although usage is accepted as an influence in fixing the pronunciation of many words, the rules of the language to which a word belongs must always be the deciding factor in most cases. In the compilation of this glossary much guidance has been obtained from the *Oxford English Dictionary, Nuttall's Standard Dictionary, Nicholson's Dictionary of Gardening, Johnson's Gardener's Dictionary* (Fraser and Hemsley), Bentham and Hooker's botanical works, Dr. B. Daydon Jackson's *Glossary of Botanical Terms*, Dr. R. T. Harvey-Gibson's *Plant Names and their Derivations*, and G. F. Zimmer's *Popular Dictionary of Botanical Terms*, among other works.

In offering this book to the public I may add in conclusion that while I claim no profound knowledge of the classics or of botany I do know something of the simple wants of my fellow amateur gardeners. It is with that conviction that I have made this effort to smooth down one of the roughest places in our common pursuit. To satisfy and to please everyone is far beyond my aspirations. But if in this work I have succeeded in reducing even by a little the menace presented us by that "pile of heterogeneous names which stand as a barrier between our people and the fairest gates of knowledge" (*Botany*, by Professor Earle), I shall feel that I have done my bit in a good cause.

1931                                                    A. T. JOHNSON

IT HAS FALLEN to my lot to tread delicately in the footsteps of the author of the original edition of this work, and to graft, more or less successfully, my work upon his. The main task entrusted to me has been to increase the scope of the work. To this end many more genera and species of plants appear than in the first edition, which was mainly limited to hardy plants, shrubs, etc. The additions therefore comprise a large number of indoor temperate and tropical subjects, while also the opportunity has been taken to include the better known names of the new sectional genera into which the cacti, mesembryanthemums and houseleeks have been grouped in modern times.

It is therefore the hope of both publisher and joint author that PLANT NAMES SIMPLIFIED may appeal in a wider measure to the new generations of both amateur and professional gardeners and have as useful a life as the first edition enjoyed in the fifteen years of its existence.

1946                                                    HENRY A. SMITH

# PLANT NAMES SIMPLIFIED

**Abelia**, a-*beel*-e-a; after Dr. Clarke Abel, physician, and author on China, who discovered A. chinensis, 1816-1817. Flowering shrubs.
> CHINENSIS, tshi-*nen*-sis, of China.
> FLORIBUNDA, flor-ib-*un*-da, many-flowered.
> GRANDIFLORA, gran-dif-*lo*-ra, large-flowered.
> TRIFLORA, trif-*lo*-ra, three-flowered, *i.e.*, flowers in threes.
> UNIFLORA, u-nif-*lo*-ra, one-flowered, *i.e.*, blooms solitary.

**Abies**, a-*be*-es (commonly *a*-beez); ancient L. name, possibly from L. *abeo*, depart, *i.e.*, from the ground, referring to great height attained by some species. Coniferous trees.
> AMABILIS, am-*a*-bil-is, lovely.
> BALSAMEA, bal-*sa*-me-a, aromatic (Balm of Gilead fir).
> BRACHYPHYLLA, brak-if-*il*-la, short-leaved.
> BRACTEATA, brak-te-*a*-ta, having bracts, or modified leaves, at bases of leaf-stalks.
> CEPHALONICA, sef-a-*lon*-ik-a, of Cephalonia.
> CONCOLOR, kon-*kol*-or, one-coloured, *i.e.*, of uniform tint.
> GRANDIS, *gran*-dis, of great size.
> MAGNIFICA, mag-*nif*-ik-a, magnificent, beautiful.
> NOBILIS, *no*-bil-is, noble, stately.
> PECTINATA, pek-tin-*a*-ta, comb-like, alluding to leaves.

**Abobra**, a-*bob*-ra; native Brazilian name. Scarlet-fruited climber of the cucumber family.
> VIRIDIFLORA, ver-id-if-*lo*-ra, green-flowered.

**Abronia**, a-*bro*-ne-a; from Gr. *abros*, delicate, alluding to leafy involucre enclosing unopened blossoms. Trailing plants.
> LATIFOLIA, lat-if-*o*-le-a, broad-leaved.
> UMBELLATA, um-bel-*la*-ta, having blossoms in umbels.

**Abrus**, *a*-brus; from Gr. *abros*, delicate, with reference to the soft leaves. Warmhouse climber.
> PRECATORIUS, prek-a-*tor*-e-us, entreating, the reference being to the black and red seeds of which rosaries are made.

**Abutilon**, a-*bu*-til-on; the Arabic name for a mallow-like plant. Greenhouse shrubs.
> DARWINII, *dar*-win-ei, after Darwin.
> INSIGNE, in-*sig*-ne, handsome.
> MEGAPOTAMICUM, meg-ap-o-*tam*-ik-um, meaning big river—the Rio Grande.
> SELLOWIANUM MARMORATUM, sel-lo-ve-*a*-num mar-mor-*a*-tum, Sellow's marbled—the foliage.
> THOMPSONII, tom-*so*-nei, after Thompson.
> VEXILLARIUM, veks-il-*lar*-e-um, standard bearing.
> VITIFOLIUM, vi-tif-*o*-le-um, vitis (vine)-like—the leaves.

**Acacia**, a-*ka*-she-a; Gr. *akazo*, to sharpen, or "akakia," the Egyptian Thorn (A. arabica), from which gum-arabic is obtained. Greenhouse flowering shrubs.
> ARMATA, ar-*ma*-ta, armed—with thorns.
> BAILEYANA, ba-le-*a*-na, after Bailey. Also sold as Mimosa.
> DEALBATA, de-al-*ba*-ta, whitened—the foliage. The Mimosa of florists' shops.
> DRUMMONDII, drum-*mon*-dei, after Drummond.
> RICEANA, rice-*a*-na, after Rice.
> VERTICILLATA, ver-tis-il-*la*-ta, whorl-leaved.

**Acaena**, ak-*e*-na (or ass-*e*-na); from Gr. *akanthos*, a thorn, many kinds having a spiny calyx. Creeping rock plants.
> BUCHANANII, bu-kan-*a*-nei, after Buchanan.
> MICROPHYLLA, mi-krof-*il*-la, small-leaved.
> NOVÆ-ZEALANDEÆ, *nov*-e-zeel-*and*-e-e, of New Zealand.

**Acalypha**, a-*kal*-y-fa; Gr. *akalepe*, Hippocrates' name for nettle. Warm-house shrubby plants with variegated foliage.
> GODSEFFIANA, god-*sef*-fe-a-na, after Joseph Godseff.
> HISPIDA, *his*-pid-a, clothed with stiff hairs.
> MACROPHYLLA, mak-rof-*il*-a, large-leaved.
> MUSAICA, mu-*sa*-ik-a, mosaic, referring to the blotched colouring of the leaves.
> SANDERI, *san*-der-i, after Messrs. Sander, nurserymen.

1

**Acanthocereus**, a-*kan*-tho-*se*-re-us; from Gr. *acanthos*, a thorn and *cereus*, a well-known genus of cacti. Greenhouse cactus.
PENTAGONUS, pent-ag-*o*-nus, having five angles—the stems.

**Acantholimon**, ak-an-*thol*-e-mon; derivation obscure, but Gr. *akanthos*, a prickle, alludes to spiny foliage. Rock plants.
GLUMACEUM, glu-*ma*-se-um, with chaffy bracts.
VENUSTUM, ven-*us*-tum, pleasing, lovely.

**Acanthus**, ak-*an*-thus; Gr. *akanthos*, a prickle, some species being spiny. A conventional form of the leaf of A. spinosus is used in architecture. Herbaceous plants.
MOLLIS, *mol*-lis, soft or tender, usually means velvety.
SPINOSUS, spi-*no*-sus, spiny.

**Acer**, *a*-ser; classical L. name, possibly from L. *acer*, hard or sharp, the wood once having been used for writing tablets. Also pronounced *ak*-er. Trees.
CAMPESTRE, kam-*pes*-tre, growing in fields. The English Maple.
DASYCARPUM, das-i-*carp*-um, hairy-fruited.
GINNALA, jin-*na*-la, vernacular name.
GLABRUM, *glabe*-rum, smooth, hairless.
INSIGNE, in-*sig*-ne, remarkable.
JAPONICUM, jap-*on*-ik-um, of Japan.
MACROPHYLLUM, mak-rof-*il*-lum, large-leaved.
PALMATUM, pal-*ma*-tum, leaves palmate, like a hand.
PENNSYLVANICUM, pen-sil-*va*-nik-um, of Pennsylvania.
PLATANOIDES, plat-an-*oy*-des, like a plane tree. The Norway Maple.
PSEUDO-PLATANUS, sued-o-plat-*a*-nus, false plane tree. The Sycamore.
RUBRUM, *roo*-brum, red-flowered.
SACCHARUM, sak-*ar*-um, sugar. The Sugar Maple.

**Aceranthus**, as-er-*an*-thus; from L. *acer*, sharp, i.e., pointed, and Gr. *anthos*, a flower. Woodland plant.
DIPHYLLUS, dif-*il*-us, two-leaved.

**Aceras**, *a*-ser-as; from Gr. *a*, without, and *keras*, a horn, the flowers having no spur. Terrestrial orchid.
ANTHROPOPHORA, an-thro-*pof*-or-a, resembling a man. The Green Man Orchis.

**Achillea**, ak-il-*e*-a; after Achilles, the Greek hero, who first used the plant in medicine. Herbaceous and rock plants.
ALPINA, al-*pine*-a, or al-*pin*-a, of the Alps or alpine.
ARGENTEA, ar-*jen*-te-a, silvery-white—the foliage.
COMPACTA, kom-*pak*-ta, compact.
MILLEFOLIUM, mil-le-*fo*-leum, thousand-leaved. The Yarrow or Milfoil of which there are cultivated forms.
MONGOLICA, mon-*gol*-ik-a, Mongolian.
MONTANA, mon-*ta*-na, of mountains.
PTARMICA, *tar*-mik-a, Gr. *ptarmos*, sneezing, dried flowers once used for snuff. The Sneezewort.
RUPESTRIS, roo-*pes*-tris, growing on rocks.
SANTOLINA, san-to-*le*-na, resembles Santolina.
SERBICA, *ser*-bik-a, of Serbia.
TOMENTOSA, to-men-*to*-sa, downy foliage.

**Achimenes**, ak-e-*mee*-neez; from L. *cheimanos*, tender, as to cold. Greenhouse herbaceous perennials. Many florists' hybrids.
CARMINATA, kar-*min*-a-ta, carmine.
COCCINEA, kok-*sin*-e-a, scarlet.
LONGIFLORA, long-if-*lo*-ra, long-flowered.

**Acineta**, ak-in-*e*-ta; from L. *akineta*, immovable, the lip being jointless. Cool-house orchid.
HUMBOLTII, hum-*bolt*-ei, after the noted traveller Humboldt.

**Acokanthera**, ak-*o*-kan-*the*-ra; from L. *akoke*, a mucron or point, and *anthera*, an anther, the anthers are mucronate. Greenhouse flowering shrubs.
SPECTABILIS, spek-*tab*-il-is, showy. The Winter Sweet.

**Aconitum**, ak-o-*ni*-tum; ancient classical name, probable origin, Gr. *akon*, a dart, arrows at one time being poisoned with the juices of the plant. Herbaceous plants.
FISCHERI, fish-*er*-i, after Fischer, a professor of botany.
FORRESTII, *for*-res-tei, after Forrest, plant collector.
FORTUNEI, for-*tune*-i, after Fortune, plant collector.
JAPONICUM, jap-*on*-ik-um, Japanese.
LYCOCTONUM, lik-*ok*-to-num, wolf's-bane.
NAPELLUS, nap-*el*-lus, turnip-rooted. The Monkshood.
VARIEGATUM, var-e-eg-*a*-tum, variegated.
VOLUBILE, vol-*u*-bil-e, twining.
WILSONII, wil-*so*-nei, after Prof. E. H. Wilson.

**Acorus,** ak-*or*-us, ancient name, possibly from Gr. *a*, without, and *kore*, pupil of the eye, alluding to ancient use of plant in medicine. Aquatics.

    CALAMUS, *kal*-a-mus; *kalon*, Arabic for reed; Calamus, L. name for the Sweet-flag.

    GRAMINEUS, gram-*in*-e-us, grass-leaved.

**Acroclinium,** ak-ro-*klin*-e-um; from Gr. *akros*, top, and *kline*, a bed or couch, referring to the centres of the flowers. Annual everlasting flowers.

    ROSEUM, *ro*-ze-um, rosy coloured.

**Acrostichum,** ak-*ros*-tik-um; from Gr. *akros*, top, and *stichos*, a row, application unknown. Tropical ferns.

    CRINITUM, *kryn*-it-um, hairy.

    PELTATUM, pel-*ta*-tum, peltate or shield-shaped—the fertile fronds.

**Actæa,** ak-*te*-a, from Gr. *aktaia*, elder, the leaves resembling those of that tree. Herbaceous plants.

    ALBA, *al*-ba, white, the berries.

    SPICATA, spe-*ka*-ta, spiked, alluding to inflorescence.

**Actinella,** ak-tin-*ell*-a; from Gr. *aktis*, a ray, flowers being rayed like small sun-flowers; lit. a little ray. Rock and border perennials.

    ACAULIS, a-*kaw*-lis, stalkless, or apparently so.

    GRANDIFLORA, gran-dif-*lo*-ra, large-flowered.

**Actinidia,** ak-tin-*id*-e-a; from Gr. *aktis*, a ray, referring to star-like flowers, or to rayed stigmas of female blooms. Climbing shrubs.

    ARGUTA, ar-*gu*-ta sharp-toothed or serrated.

    CHINENSIS, tshi-*nen*-sis, of China.

    HENRYI, *hen*-re-i after Dr. A. Henry.

    VOLUBILIS, vol-*u*-bil-is, twining.

**Actinotus,** ak-tin-*o*-tus; from L. *actinotos*, rayed, referring to the involucre. Tender perennial.

    HELIANTHI, he-le-*anth*-i, sun-flower-like—the petal-like involucre. The Australian Flannel Flower.

**Ada,** *a*-da, a complimentary name. Greenhouse orchid.

    AURANTIACA, aw-ran-te-*a*-ka, orange coloured.

**Adenophora,** ad-en-*of*-or-a; from Gr. *aden*, a gland, and *phoreo*, to bear; reference obscure. Herbaceous plants.

    DENTICULATA, den-tik-u-*la*-ta, finely-toothed.

**Adenophora** (*continued*)

    LATIFOLIA, lat-if-*o*-le-a, broad-leaved.

    LILIIFLORA, lil-i-if-*lo*-ra, lily-flowered.

    POLYMORPHA, pol-i-*mor*-fa, many forms, *i.e.*, of same parts.

    STYLOSA, sty-*lo*-sa, long-styled.

    VERTICILLATA, ver-tis-il-*la*-ta, whorl-leaved.

**Adiantum,** ad-e-*an*-tum; from Gr. *a*, not, and *diantos*, moistened (*adiantos*, dry), the fronds of the Maidenhair Fern being supposed to remain dry even after being plunged under water. Greenhouse and hardy ferns.

    AFFINE, af-*fin*-e, related.

    CAUDATUM, kaw-*da*-tum, tailed.

    CAPILLUS-VENERIS, kap-*il*-lus *ven*-er-is, Venus's hair. The British Maidenhair fern.

    CONCINNUM, kon-*sin*-num, neat.

    C. LATUM, *la*-tum, broad—*i.e.*, broader than the type.

    CUNEATUM, ku-ne-*a*-tum, wedge-shaped—the pinnæ. The popular Maidenhair fern.

    DECORUM, dek-*or*-um, decorous—shapely or becoming.

    FARLEYENSE, far-ley-*en*-se, of Farley Hill, Barbados, where it originated.

    FORMOSUM, for-*mo*-sum, beautiful.

    FULVUM, *ful*-vum, tawny.

    GRACILLIMUM, gra-*sil*-lim-um, most graceful.

    MACROPHYLLUM, mak-*rof*-il-lum, large fronds—the size of the pinnæ.

    PEDATUM, ped-*a*-tum, like a bird's foot—the fronds.

    PRINCEPS, *prin*-seps, princely.

    RENIFORME, ren-e-*for*-me, kidney-shaped—the fronds.

    TENERUM, *ten*-er-um, tender.

    TETRAPHYLLUM, tet-raf-*il*-lum, four-leaved, *i.e.*, pinnæ.

    TRAPEZIFORME, trap-e-zif-*or*-me, rhomb-leaved—the pinnæ.

    WILLIAMSII, *will*-yams-ei, after B. S. Williams, nurseryman, of Holloway.

**Adlumia,** ad-*lu*-me-a; named after Major John Adlum, an American author. Biennial climber.

    CIRRHOSA, sir-*ro*-sa, tendrilled.

**Adonis,** ad-*o*-nis; after Adonis of the classics, one of Venus's lovers, whose blood is supposed to have stained the petals of the Pheasant-eye Adonis (A. autumnalis). Annuals and perennials.

    ÆSTIVALIS, es-tiv-*ale*-is, of summer—time of flowering.

    AMURENSIS, am-ure-*en*-sis, of Amur.

**Adonis** (*continued*)

AUTUMNALIS, au-tum-*na*-lis, of autumn—time of flowering.

PYRENAICA, pir-en-*a*-ik-a, Pyrenean.

VERNALIS, ver-*na*-lis, of spring—time of flowering.

**Æchmea**, *eek*-me-a; from Gr. *aichme*, a point; referring to the rigid points on the floral envelopes in the bud stage. Warm house herbaceous plants.

CŒLESTIS, se-*les*-tis, sky blue.

FASCIATA, fas-*si*-a-ta, banded.

FULGENS, *ful*-jenz, glowing red.

MARIÆ REGINÆ, mar-e-e re-*ji*-ne, after Queen Maria.

**Aeonium**, a-*o*-ne-um; ancient name used by Dioscorides for a plant similar to A. arboreum. Greenhouse succulents.

ARBOREUM, ar-bor-*e*-um, tree-like.

CÆSPITOSUM, ses-pit-*o*-sum, tufted.

CANARIENSE, ka-nar-e-*en*-se, of the Canary Islands.

HAWORTHII, ha-*worth*-ei, after Adrian Hardy Haworth, author of literature on succulents.

HOLOCHRYSUM, hol-o-*kry*-sum, entirely yellow—the flowers.

SPATHULATUM, spath-ul-*a*-tum, from spatula, a spoon—the shape of the leaves; this species is not frequent but a hybrid between it and cæspitosum is common in greenhouses and windows.

TABULÆFORME, tab-ul-e-*for*-me, table-like, the rosettes have a flat top.

**Aerides**, a-er-*i*-deez; from L. *aer*, the air, some of the species obtain all their nourishment from the atmosphere through aerial roots. Tropical orchids.

CRISPUM, *kris*-pum, curled.

ODORATUM, od-o-*ra*-tum, sweet smelling.

**Æschynanthus**, ees-kin-*anth*-us; from Gr. *aischuno*, to be ashamed, and *anthos*, a flower, referring to the modest flowers of some species. Warm house shrubby and pendent plants.

FULGENS, *ful*-jenz, glowing red.

PULCHER, pul-ker, beautiful.

SPECIOSUS, spes-e-*o*-sus, handsome.

**Æsculus**, *es*-ku-lus; ancient L. name of an oak or mast-bearing tree. Flowering trees.

CALIFORNICA, kal-if-*orn*-i-ka, Californian.

CAPPADOCICA, kap-pa-*do*-se-ka, Cappadocian.

CARNEA, *kar*-nea, flesh-coloured.

HIPPOCASTANUM, hip-po-*kas*-ta-num, Gr. *hippos*, a horse, and L. *castanea*, the

**Æsculus** (*continued*)

Chestnut tree of Virgil. The Horse Chestnut.

MACROSTACHYA, mak-ros-*tak*-e-a, large spiked.

PARVIFLORA, par-vif-*lo*-ra, small-flowered.

RUBICUNDA, roo-be-*kun*-da, red-flowered.

**Æthionema**, eth-e-*o*-ne-ma; origin obscure, said to be derived from Gr. *aitho*, to burn, and *nema*, a filament, alluding to burnt appearance of stamens; more probable origin, the burning or acrid taste of some species. Rock plants.

ARMENUM, ar-*me*-num, of Armenia.

GRANDIFLORUM, gran-dif-*lo*-rum, large-flowered.

IBERIDEUM, i-ber-*id*-e-um, like an Iberis (Candytuft).

PULCHELLUM, pul-*kel*-lum, beautiful.

**Agapanthus**, ag-a-*pan*-thus; from Gr. *agape*, love, and *anthos*, a flower. Greenhouse herbaceous plants.

UMBELLATUS, um-bel-*la*-tus, flowers in umbels. The Blue African Lily.

**Agaricus**, ag-*ar*-ik-us; probably from Agari, a district in Sarmatia. Fungi, including the edible Mushroom.

CAMPESTRIS, kam-*pes*-tris, growing in fields or plains. The Common Field Mushroom.

**Agathæa**, ag-a-*the*-a; from Gr. *agathos*, good, alluding to beauty of the flower. Greenhouse sub-shrub.

CŒLESTIS, se-*les*-tis, heavenly, sky blue.

**Agave**, ag-*ah*-vee; from Gr. *aganos*, noble, referring to stately form when in flower. Greenhouse succulents.

AMERICANA, a-mer-ik-*a*-na, of South America. The American Aloe (of gardens).

ATROVIRENS, a-tro-*ver*-enz, dark green.

ATTENUATA, at-ten-u-*a*-ta, attenuated or drawn out.

HORRIDA, *hor*-rid-a, horrid, having strong spines.

UNIVITTATA, u-ne-vit-*ta*-ta, one line or stripe—on the leaves.

UTAHENSIS, *u*-tah-*en*-sis, of Utah.

**Ageratum**, aj-er-*a*-tum; from Gr. *a*, not, and *geras*, old, or Gr. *ageratos*, not growing old, presumably meaning that the flowers do not readily assume a withered appearance. Summer bedder.

CONYZOIDES, kon-iz-*oy*-dez, Conyza-like.

MEXICANUM, mex-ik-*a*-num, of Mexico.

4

**Aglaonema**, ag-la-on-*e*-ma; from Gr. for bright thread, possibly referring to stamens. Tropical aroids.

COSTATUM, kos-*ta*-tum, leaves strongly ribbed.

PICTUM, *pik*-tum, painted—the blotched leaves.

**Agrostemma**, ag-ro-*stem*-ma; from Gr. *agros*, a field, and *stemma*, a crown or wreath, referring to the beauty of the flowers. This genus is now joined with Lychnis, which see.

**Agrostis**, a-*gros*-tis; from Gr. *agros*, a field, the Greek name for a grass. Annual ornamental grasses.

NEBULOSA, neb-ul-*o*-sa, nebulous or cloud-like—the inflorescence. The Cloud Grass.

PULCHELLA, pul-*kel*-la, pretty.

**Aichryson**, a-*kry*-son; ancient name used by Dioscorides for a plant similar to Aeonium (Sempervivum) arboreum. Greenhouse succulents.

DOMESTICUM FOLIIS VARIEGATIS, do-*mes*-tik-um *fol*-e-is var-e-eg-*a*-tis, a variegated plant common in greenhouses and windows.

DICHOTOMUM, di-kot-*o*-mum, stem divides into branches of equal size.

DIVARICATUM, de-var-ik-*a*-tum, wide spreading branches.

**Ailanthus**, a-*lan*-thus; from *ailanto*, the native (Chinese) name for one of the species, and signifying tall enough to reach the skies. Deciduous tree.

GLANDULOSA, glan-dul-*o*-sa, glandular—the leaflets. The Tree of Heaven.

**Ajuga**, a-*ju*-ga; a corruption of L. *abiga*, a plant used in medicine, or (more probably) from *a*, no, and *zugon*, a yoke, in reference to the calyx lobes being equal—not bilabiate. Creeping plants.

GENEVENSIS, jen-e-*ven*-sis, of Geneva.

REPTANS, *rep*-tans, creeping. Several forms of this are cultivated.

**Akebia**, ak-*e*-be-a; an adaptation of the Japanese name for these shrubby twining plants.

LOBATA, la-*ba*-ta, lobed, *i.e.*, the leaves.

QUINATA, kwin-*a*-ta, five-lobed—the leaves.

**Albuca**, al-*bu*-ka; from L. *albicans* or *albus*, white—the prevailing colour. Greenhouse bulbous plants.

NELSONII, nel-*so*-nei, after Nelson.

**Alchemilla**, al-kem-*il*-la; from Arabic *alkemelych*, alluding to use of plants in alchemy. Rock plants.

ALPINA, al-*pine*-a, alpine.

SERICEA, ser-*is*-e-a, silky—the leaves.

VULGARIS, vul-*gar*-is, common. The Lady's Mantle.

**Alisma**, al-*iz*-ma; derivation doubtful. Aquatic or bog plants.

NATANS, *na*-tanz, floating.

PLANTAGO, plan-*ta*-go, an old generic name referring to the Plantago (plantain)-like leaves. The Water Plantain.

**Allamanda**, al-la-*man*-da; after Dr. Allamand, professor of natural history at Leyden. Warm-house shrubby climbers.

HENDERSONII, hen-der-*so*-nei, after Henderson.

NOBILIS, *no*-bil-is, large or noble.

SCHOTTII, *shot*-tei, after Schott.

**Allium**, *al*-le-um; the Latin term for Garlic; now the name of all the onion family, or from Celtic *all*, meaning pungent or burning. Bulbous perennials and culinary herbs.

ASCALONICUM, as-kal-*o*-nik-um, of Askelon, Palestine. The Shallot or Eschallot.

BEESIANUM, bees-e-*a*-num, introduced by Messrs. Bees.

CEPA, *ke*-pa, headed, probably from Celtic *cep*, a head. The Onion.

CHRYSANTHUM, kris-*an*-thum, golden-flowered.

CYANEUM, sy-*a*-ne-um, blue-flowered.

DESCENDENS, de-*sen*-dens, the colour of the head of blooms commencing at the apex and spreading downwards.

FISTULOSUM, fis-tu-*lo*-sum, fistular or hollow leaved. Welsh or Ciboul Onion.

FLAVIDUM, *fla*-vid-um, pale yellow.

GIGANTEUM, ji-*gan*-te-um, gigantic.

KANSUENSE, kan-su-*en*-se, from Kansu.

MOLY, *mo*-le, old Gr. name.

NARCISSIFLORUM, nar-sis-if-*lor*-um, Narcissus-flowered.

NEAPOLITANUM, ne-ap-ol-e-*ta*-num, from Naples.

OSTROWSKYANUM, os-trow-ske-*a*-num, after Ostrowsky, a Russian botanist.

PANICULATUM, pan-ik-u-*la*-tum, panicled.

POLYPHYLLUM, pol-if-*il*-lum, many-leaved.

PORRUM, *por*-rum, from Celtic *pori*, to eat. The Leek.

PULCHELLUM, pul-*kel*-lum, beautiful but small.

**Allium** (*continued*)

PURDOMII, *pur*-dom-ei, after Purdom.
ROSEUM, *ro*-ze-um, rosy.
SATIVUM, *sat*-iv-um, cultivated. The Garlic.
SCHŒNOPRASUM, sken-*op*-ras-um, old Greek name for Leek. The Chives.
SCORODOPRASUM, scor-od-*op*-ras-um, combination of Greek for onion and leek, signifying the plant is both onion and leek. The Rocambole or Sand Leek.
SIKKIMENSE, sik-kim-*en*-se, from Sikkim.
SPHÆROCEPHALUM, sfer-o-*sef*-al-um, old generic name meaning round-headed —the flower head.
TRIQUETRUM, tri-*kwet*-rum, three-cornered—the stalks.

**Alnus**, *al*-nus; *alnus*, Latin name for alder. Trees.

CORDATA, kor-*da*-ta, heart-shaped—the leaves.
GLUTINOSA, glu-tin-*o*-sa, sticky—the foliage gummy. The Alder.
INCANA, in-*ka*-na, grey or hoary-leaved.

**Alocasia**, al-o-*kas*-e-a; from L. *a*, without, and *colocasia*, similar to, without being, Colocasia. Tropical ornamental-leaved plants.

JENNINGSII, jen-*nings*-ei, after Jennings.
MACRORHIZA VARIEGATA, mac-rorh-*i*-za var-e-eg-*a*-ta, long-rooted and variegated leaved.
METALLICA, met-*al*-lik-a, of metallic colouring.
SANDERIANA, san-der-e-*a*-na, after Messrs. Sander of St. Albans, nurserymen.
THIBAUTIANA, the-baut-e-*an*-a, after Thibaut.
WATSONIANA, wat-*so*-ne-an-a, after W. Watson, of Kew.

**Aloe**, *al*-o-e; old Arabic name, possibly from Arabic *alloch*, referring to species used medicinally. Greenhouse succulent plants. English name, Aloe, al-o.

ABYSSINICA, ab-is-*in*-ik-a, Abyssinian.
ARBORESCENS, ar-bor-*es*-cenz, tree-like.
HUMILIS, *hum*-il-is, dwarf, humble as to stature.
MITRÆFORMIS, mit-re-*for*-mis, mitre-shaped.
SUCCOTRINA, suc-cot-*re*-na, of Socotra.
VARIEGATA, var-e-eg-*a*-ta, variegated. Partridge-breasted Aloe or Mackerel Plant.
VERA, *ve*-ra, the true or typical species.

**Alonsoa**, al-on-*so*-a; after Z. Alonzo, a Spaniard. Small half-hardy shrubby plants.

INCISIFOLIA, in-sis-if-*ol*-i-a, cut-leaved.
LINEARIS, lin-e-*ar*-is, linear-leaved.
LINIFOLIA, lin-if-*o*-le-a, Linum or flax-leaved.
WARSCEWICZI, var-skew-*ik*-zi, after Warscewicz.

**Aloysia**, al-*oy*-se-a; called after Maria Louisa, Queen of Spain. Half-hardy shrub.

CITRIODORA, sit-re-od-*or*-a, lemon-scented. The Lemon-scented Verbena.

**Alpinia**, al-*pin*-e-a; after P. Alpinus, an Italian botanist. Tropical herbaceous perennials.

NUTANS, nu-tanz, drooping or nodding.
SANDERIANA, san-der-e-*a*-na, of Messrs. Sander, St. Albans.
VITTATA, vit-*ta*-ta, striped.

**Alsine**, al-*seen*-e; Gr. *alsos*, a grove, often the habitat of the chickweeds, according to some authorities; more probably derived from Greek name given by Dioscorides to a Cerastium. Rock plants.

LARICIFOLIA, lar-is-if-*o*-le-a, Larix (larch)-like leaves.

**Alsophila**, al-*sof*-il-a; Gr. *alsos*, a grove, and *phileo*, to love, shade-loving tree ferns. Greenhouse ferns.

AUSTRALIS, aws-*tra*-lis, southern.

**Alstromeria**, al-stro-*meer*-i-a; named in honour of Baron Alstromer, a Swedish botanist. Half-hardy herbaceous plants.

AURANTIACA, aw-ran-te-*a*-ka, golden or orange.
CHILENSIS, chil-*en*-sis, of Chile.
PELEGRINA, pel-e-*gree*-na, spotted blooms.

**Alternanthera**, al-tern-*an*-ther-a; alternate anthers, alluding to the anthers being alternately barren and fertile. Dwarf, tropical, coloured-leaved plants, used for "carpet" bedding.

AMABILIS, am-*a*-bil-is, lovely; several varieties.
PARONYCHIOIDES, par-on-ik-*oy*-des, Paronychia-like; several varieties.

**Althæa**, al-*the*-a; Gr. *althaia*, a healing medium, referring to its use in medicine. Hardy herbaceous and biennial.

FICIFOLIA, fi-kif (or sif)-*ol*-e-a, Ficus or fig-leaved.

**Althæa** (*continued*)

OFFICINALIS, of-fis-in-*a*-lis, of the shop (apothecary's), applied to plants always kept "in stock" by herbalists.
ROSEA, *ro*-ze-a, red or rosy. The Hollyhock.

**Alyssum,** *al*-iss-um; from Gr. *a*, not, *lyssa*, madness, the plant once being considered a remedy for a bite by a mad dog, hence popular name, Madwort. Rock plants.

ARGENTEUM, ar-*jen*-te-um, silvery.
CORYMBOSUM, kor-im-*bo*-sum, flowers in corymbs.
MARITIMUM, mar-*it*-im-um, of the sea coast. The Sweet Alyssum.
MONTANUM, mon-*ta*-num, of mountains.
PYRENAICUM, pir-en-*a*-ik-um, Pyrenean.
SAXATILE, saks-*a*-til-e, haunting rocks.
SERPYLLIFOLIUM, ser - pil-if - *o* - le - um, thyme-leaved.

**Amarantus,** am-a-*ran*-tus; from Gr. *a*, not, and *maraino*, to fade, or *amarantos*, unfading, referring to durability or lasting quality of flowers of some species. Tender annuals. Also spelled Amaranthus.

CAUDATUS, kaw-*da*-tus, tailed—shape of inflorescence. The Love-lies-Bleeding.
CRUENTUS, kru-*en*-tus, dark blood-red.
HYPOCHONDRIACUS, hy-pok-*on*-dre-ak-us, sombre. The Prince's Feather.
TRICOLOR, *trik*-o-lor, three-coloured.

**Amaryllis,** am-a-*ril*-is; a classical name, after a shepherdess in Theocritus and Virgil, Greek and Latin poets. Half-hardy bulb.

BELLADONNA, bel-la-*don*-na, Ital. *bella*, pretty, *donna*, lady, an extract from this plant being used to brighten the eyes. The Belladonna Lily.

**Amasonia,** am-as-*o*-ne-a; after Thomas Amason, an American traveller. Tropical coloured-foliage plants.

CALYCINA, kal-ik-*een*-a, with showy calyces.
PUNICEA, pu-*nik*- (or *nis*-) e-a, reddish-purple.

**Amelanchier,** am-el-*an*-ke-er; name adapted from Fr. *amelancier*, an old name in Savoy for A. vulgaris, the Snowy Mespilus. Small flowering trees.

ALNIFOLIA, al-nif-*o*-le-a, alder-leaved.
CANADENSIS, kan-a-*den*-sis, of Canada.
VULGARIS, vul-*gar*-is, common. The Snowy Mespilus.

**Ammobium,** am-*mo*-be-um; from Gr. *ammos*, sand, and *bio*, to live, *i.e.*, thriving in sandy places. Perennial.

ALATUM, al-*a*-tum, winged—the stems.

**Ammogeton,** am-mo-*ge*-ton; from Gr. *ammos*, sand, and *geton*, alluding to natural habitat. Hardy herbaceous.

SCORZONERIFOLIUM, skor-zo-ner-if-*o*-le-um, leaves like those of Scorzonera.

**Amorpha,** am-*or*-fa; from Gr. *a*, not, and *morphe*, form, referring to irregular shape of the leaves. Shrubs.

CANESCENS, kan-*es*-ens, grey or hoary.
FRUTICOSA, frut-ik-*o*-sa, shrubby.
NANA, *na*-na, dwarf.

**Ampelopsis,** am-pel-*op*-sis, from Gr. *ampelos*, a vine, and *opsis*, resemblance; resembling a grape vine (Vitis).

QUINQUEFOLIA, kwin-ke-*fo*-le-a, five-leaved—leaves with five divisions.
TRICUSPIDATA, trik-us-pid-*a*-ta, three-pointed, the older or larger leaves with three tailed divisions.
VEITCHII, *veech*-ei, after James Veitch.

**Amsonia,** am-*so*-ne-a; after Charles Amson, a scientific American explorer. Herbaceous perennials.

ANGUSTIFOLIA, an-gus-tif-*o*-le-a, narrow-leaved.
SALICIFOLIA, sal-is-if-*o*-le-a, Salix (willow)-leaved.
TABERNÆMONTANA, tab-er-ne-mon-*ta*-na, after J. T. Tabernæmontanus, a German herbalist.

**Amygdalus,** am-*ig*-da-lus; from Gr. *amygdalos*, an almond, or from Gr. *amysso*, to lacerate, in reference to the fissures or channels in the "stone" or seed. Some believe name to be derived from Hebrew word signifying "vigilant," referring to the Almond's early blooming. Trees and shrubs.

COMMUNIS, kom-*mu*-nis, gregarious. The Almond. A.c. amara (a-*mar*-a) is the bitter almond; A.c. dulcis (*dul*-sis) is the sweet almond.

**Anacharis,** a-*nak*-ar-is; from L. and Gr. words meaning loved by ducks, being sought by aquatic fowl for food. Submerged aquatic. Syn. Elodea.

ALSINASTRUM, al-sin-*as*-trum, Alsine-like.

**Anagallis,** a-na-*gal*-is; from Gr. word meaning delightful, or possibly from a fable ascribing to the Pimpernel the power to alleviate melancholy.

**Anagallis** (*continued*)

ARVENSIS, ar-*ven*-sis, of the fields. The Pimpernel.

INDICA, *in*-dik-a, of India.

LINIFOLIA, lin-if-*o*-le-a, leaves like Linum (Flax).

MONELLII, mo-*nel*-ei, after Monell.

TENELLA, ten-*el*-la, somewhat delicate, *i.e.*, frail.

**Ananas**, an-*a*-nas (or an-*an*-as); from *nanas*, the South American name for the Pineapple. Tropical fruiting plant.

SATIVA, *sat*-iv-a, cultivated. The Pineapple.

**Ananassa**, an-a-*nas*-sa or an-an-*as*-sa, a synonym of Ananas, which see.

**Anaphalis**, an-*af*-a-lis; old Gr. name: De Candolle said the name was an ancient Gr. one for a similar plant. Herbaceous perennial.

MARGARITACEA, mar-gar-it-*a*-se-a, pearly. The Pearly Everlasting.

**Anastatica**, an-as-*tat*-ik-a; from Gr. *anastasis*, resurrection. Called Resurrection Plant because the dry dead plants open flat when immersed in water. Supposed to be the "rolling thing before the whirlwind" (Isaiah xvii, 13). Annual.

HIEROCHUNTINA, hy-er-*ok*-unt-e-na, of Jericho—the L. name of Jericho.

**Anchusa**, an-*ku*-za; from Gr. *anchousa*, a cosmetic paint for staining the skin, formerly made from A. tinctoria. Herbaceous and biennial.

CAPENSIS, ka-*pen*-sis, of the Cape of Good Hope.

ITALICA, it-*al*-ik-a, of Italy.

MYOSOTIDIFLORA, my-o-*so*-tid-if-*lor*-a, flowers like Myosotis (Forget-me-not).

OFFICINALIS, of-fis-in-*a*-lis, of the shop.

SEMPERVIRENS, sem-per-*vir*-enz, always green.

**Andromeda**, an-*drom*-ed-a; named after the Grecian Princess who was bound to a rock and rescued by the hero Perseus. Low evergreen flowering shrubs. Bog Rosemary.

FLORIBUNDA, flor-ib-*un*-da, abundant or free flowering.

JAPONICA, jap-*on*-ik-a, of Japan.

POLIFOLIA, pol-if-*o*-le-a, smooth or polished—the leaves.

**Androsace**, an-dro-*sa*-se (or an-*dro*-sa-se); from Gr. *andros*, male, and *sakos*, buckler, the anther being supposed to resemble an ancient buckler. Rock plants.

CHAMÆJASME, kam-e-*jas*-me, literally, dwarf jasmine.

CILIATA, sil-e-*a*-ta, an eyelash, *i.e.*, fringed with hairs.

FILIFORMIS, fil-if-*or*-mis, thread-like.

FOLIOSA, fo-le-*o*-sa, leafy.

GLACIALIS, glas-e-*a*-lis, of glaciers, *i.e.*, a high alpine.

HELVETICA, hel-*vet*-ik-a, of Helvetia (Switzerland).

LACTEA, lak-*te*-a, milky (white).

LACTIFLORA, lak-tif-*lo*-ra, milk-white-flowered.

LANUGINOSA, lan-u-jin-*o*-sa, with long woolly hairs.

PRIMULOIDES, prim-ul-*oy*-des, primrose-like.

PUBESCENS, pu-*bes*-enz, clothed with soft hair.

SARMENTOSA, sar-men-*to*-sa, twiggy, *i.e.*, many runners.

SEMPERVIVOIDES, sem-per-viv-*oy*-des, like a sempervivum.

SEPTENTRIONALIS, sep-ten-tre-o-*na*-lis, northern; der. *septen*, seven, *triones*, oxen, *i.e.*, the stars of the Great Bear, close to North Star, hence northern.

TIBETICA, tib-*et*-ik-a, of Tibet.

VILLOSA, vil-*o*-sa, hairy.

**Anemia**, an-*ee*-me-a; from Gr. *aneimon*, naked; refers to the naked panicles of fructification. Tropical so-called "flowering" ferns.

PHYLLITIDIS, fil-*lit*-id-is, like Phyllitis.

**Anemone**, an-*em*-o-ne; from Gr. *anemos*, wind, and *mone*, a habitation, some species enjoying windy places, hence Windflower, the English name. Herbaceous and tuberous perennials.

ANGULOSA, an-gu-*lo*-sa, angular.

APENNINA, ap-en-*ni*-na, of the Apennines.

BALDENSIS, ball-*den*-sis, of Mt. Baldo.

BLANDA, *blan*-da, enchanting or pleasing.

CORONARIA, kor-on-*a*-re-a, crown or wreath-like.

FULGENS, *ful*-jens, glowing.

HALLERI, *hal*-er-i, after Haller, a botanist.

HEPATICA, hep-*at*-ik-a, liver-like, *i.e.*, the lobed leaves.

HORTENSIS, hor-*ten*-sis, of gardens.

JAPONICA, jap-*on*-ik-a, of Japan.

NARCISSIFLORA, nar-sis-if-*lo*-ra, narcissus-flowered.

NEMOROSA, nem-or-*o*-sa, of open glades. The Wood Anemone.

**Anemone** (*continued*)

PATENS, *pa*-tens, spreading open, or standing out.

PULSATILLA, pul-sat-*il*-la, to shake, *i.e.*, in the wind.

RANUNCULOIDES, ra-nun-kul-*oy*-des, like a Ranunculus (Buttercup).

RIVULARIS, riv-u-*lar*-is, of brooks or streams.

RUPICOLA, ru-*pik*-o-la, a rock-dweller.

SULPHUREA, sul-few-*re*-a, sulphur-coloured.

SYLVESTRIS, sil-*ves*-tris, pertaining to woods.

VERNALIS, ver-*na*-lis, of the spring.

**Anemonopsis,** an-em-on-*op*-sis; from *anemone*, and Gr. *opsis*, a resemblance, referring to the flowers. Herbaceous plant.

MACROPHYLLA, mak-rof-*il*-la, long-leaved.

**Angelica,** an-*jel*-ik-a; from L. *angelus*, an angel, or angelic, alluding to valuable healing properties. Waterside perennials.

HIRSUTA, hir-*su*-ta, hairy.

OFFICINALIS, of-fis-in-*a*-lis, of apothecary's shop. Commercial Angelica.

**Angelonia,** an-gel-*o*-ne-a; from *angelon*, its South American name. Tropical herbaceous perennials.

GRANDIFLORA, gran-dif-*lo*-ra, large flowered.

**Angræcum,** an-*gra*-kum; L. form of Angrek, the Malay name for all orchids of this habit. Tropical epiphytal orchids.

EBURNEUM, eb-*ur*-ne-um, like ivory.

SESQUIPEDALE, ses-kwip-ed-*a*-le, a foot-and-a-half, the reference being to the long floral spur or nectary.

**Anguloa,** ang-ul-*o*-a; after Angulo, a Spanish naturalist. South American orchids.

CLOWESII, *klows*-ei, after Clowes.

**Anhalonium,** an-hal-*o*-ni-um; possibly from L. *an*, without, and *helos*, nail or spike, in allusion to absence of spines or thorns. Greenhouse cacti.

FISSURATUM, fis-sur-*a*-tum, cleft.

PRISMATICUM, priz-*mat*-ik-um, cut like a prism.

**Anomatheca,** an-o-math-*e*-ka; from Gr. *anomos*, singular (irregular), and *theca*, a capsule, referring to the form of the seed-pod. Bulbs.

CRUENTA, kroo-*en*-ta, blood-red.

**Antennaria,** an-ten-*na*-re-a; from L. *antenna*, a sail-yard, the hairs attached to seed of the plant resembling the antennæ (feelers) of insects. Antennæ is of the same derivation. Rock and border plants.

CARPATICA, kar-*pat*-ik-a, Carpathian.

DIOICA, di-*oy*-ka, lit. two houses, *i.e.*, male and female parts being in separate flowers (diœcious).

PLANTAGINIFOLIA, plan-ta-jin-if-*o*-le-a, Plantago (plantain)-leaved.

**Anthemis,** *an*-the-mis; Gr. name for chamomile. Annual, biennial and perennial herbs with strongly scented foliage.

BIEBERSTEINII, bi-ber-*sti*-nei, after Bieberstein, a Russian botanist.

COTULA, *kot*-u-la, cup-like, presumably the flower-head.

MACEDONICA, mas-e-*don*-ik-a, of Macedonia.

NOBILIS, *no*-bil-is, noble—large flowers for a small plant. The Chamomile.

TINCTORIA, tink-*tor*-e-a, of dyers, *tingo*, to dye. Dyer's Chamomile.

**Anthericum,** an-*ther*-ik-um; from Gr. *anthos*, a flower, and *kerkos*, a hedge, probably alluding to great height of some species. Bulbous plants.

LILIAGO, lil-e-*a*-go, the silvery (St. Bernard's) lily.

LILIASTRUM, lil-e-*as*-trum, the star (St. Bruno's) lily.

RAMOSUM, ram-*o*-sum, with many branches.

VARIEGATUM, var-e-eg-*a*-tum, variegated leaves.

YEDOENSE, yed-o-*en*-se, of Yeddo, in Japan.

**Antholyza,** an-thol-*i*-za; from Gr. *anthos*, a flower, and *lyssa*, rage; a metaphorical name, the opening blossoms being supposed to resemble the mouth of an enraged animal. Bulbous (corm) plants.

PANICULATA, pan-ik-ul-*a*-ta, flowers in a panicle or branching inflorescence.

**Anthriscus,** an-*ihris*-kus; the Gr. name of a similar plant described by Pliny. Culinary herb.

CEREFOLIUM, *ke*-ref-o-le-um (or *se*-ref-o-le-um), waxy. The Chervil.

**Anthurium,** an-*thu*-re-um; from Gr. *anthos*, a flower, and *oura*, a tail, alluding to the spadix. Tropical her-

**Anthurium** (*continued*)
baceous plants, either fine foliage or floral.
ANDREANUM, an-dre-*an*-um, after André.
CRYSTALLINUM, kris-tal-*leen*-um, crystalline—as to the veining.
SCHERZERIANUM, sher-zer-e-*a*-num, after Scherzer.
VEITCHII, *veech*-ei, after James Veitch.
WAROQUEANUM, war-ok-e-*a*-num, after Waroque.

**Anthyllis,** an-*thil*-is; from Gr. *anthos*, a flower, and *ioulos*, down, the calyx in many species being downy. Shrubs and perennials.
BARBA-JOVIS, *bar*-ba-*jo*-vis, Jupiter's or Jove's Beard.
HERMANNIÆ, her-*man*-e-e, after Frau Hermann.
SERICEA, ser-*is*-e-a, silky.
TETRAPHYLLA, tet-raf-*il*-a, four-leaved.
VULNERARIA, vul-ner-*a*-re-a, wound-healing.

**Antirrhinum,** an-ter-*rhi*-num; from Gr. *anti*, resembling, and *rhis* (*rhinos*), a snout, alluding to the shape of the flower. Border and rock plants.
ASARINA, as-ar-*e*-na, Asarrhina, an older generic name for the plant; probable meaning, gummy-snouted.
MAJUS, *ma*-jus, great. The Snapdragon.

**Aotus,** a-*o*-tus; from *a*, without, and *ous*, an ear, certain calyx appendages are wanting that are present in an allied genus—Pultenæa. Greenhouse evergreen shrubs.
GRACILLIMA, gra-*sil*-lim-a, most slender.

**Aphelandra,** af-el-*an*-dra; from Gr. *apheles*, simple, and *andros*, male, the anthers being one-celled. Tropical evergreen flowering shrubs.
AURANTIACA, aw-ran-te-*a*-ka, golden-orange.
NITENS, *nit*-enz, shining.
ROEZLII, ro-*ez*-lei, after Roezl.

**Aphyllanthes,** af-il-*an*-thes; from Gr. *a*, without, and *phyllon*, a leaf, the flowers being borne at the tips of rush-like growths. Herbaceous perennial.
MONSPELIENSIS, mon-spe-li-*en*-sis, of Montpelier.

**Apium,** *ap*-i-um; from *apon*, Celtic for water. Ditch plants and culinary vegetables.
GRAVEOLENS, *grav*-e-ol-ens, strong-smelling. The Celery.
G. RAPACEUM, rap-*a*-se-um, turnip-like. The Turnip-rooted Celery or Celeriac.

**Aporocactus,** a-por-o-*kak*-tus; from Gr. *aporos*, not open or impenetrable, and cactus—no special application suggested.
FLAGELLIFORMIS, flaj-el-lif-*or*-mis, flagellant or whip-like. The Rat's-tail Cactus.
MALLISONII, mal-le-*son*-ei, after Mallison.

**Aptenia,** ap-*ten*-e-a, from Gr. *apten*, wingless, in allusion to the valves of the capsule having no wings. Greenhouse succulent.
CORDIFOLIA VARIEGATA, kor-dif-*o*-le-a var-e-eg-*a*-ta, variegated heart-shaped leaves. The popular golden-leaved bedding Mesembryanthemum.

**Aponogeton,** a-pon-o-*ge*-ton; from Celtic, *apon*, water, and *geiton*, neighbour; or Gr. *apo*, away from, and *ge*, the earth, *i.e.*, living in water. Floating aquatic.
DISTACHYON, dis-*tak*-e-on, two-spiked—the V-shaped flower spike.

**Aquilegia,** ak-wil-*e*-je-a; origin doubtful, possibly from L. *aquila*, an eagle, the flower spur resembling an eagle's claw; Eng. name, Columbine, from L. *columba*, a dove, the form of the flower suggesting a group of doves. Herbaceous perennials; many hybrid strains.
CÆRULEA, se-*ru*-le-a, dark blue.
CANADENSIS, kan-a-*den*-sis, Canadian.
CHRYSANTHA, kris-*an*-tha, golden-flowered.
FLABELLATA, flab-el-*la*-ta, fan-shaped.
GLANDULOSA, glan-dul-*o*-sa, glandular.
GLAUCA, *glaw*-ka, bluish-grey, glaucous.
JUCUNDA, juk-*un*-da, bright (joyous).
KITAIBELII, kit-a-*bel*-ei, after P. Kitaibel, a professor of botany.
NEVADENSIS, nev-a-*den*-sis, of Nevada.
REUTERI, *roy*-ter-i, after G. Reuter, botanist and collector.
SKINNERI, *skin*-er-i, after Skinner, a botanist.
THALICTRIFOLIA, thal-ik-trif-*o*-le-a, leaved like a Thalictrum.
VIRIDIFLORA, vir-id-if-*lo*-ra, green flowered.
VULGARIS, vul-*gar*-is, common. The Columbine.

**Arabis,** *ar*-ab-is; from Gr. *arabis*, Arabia, the home of several species. Rock plants.
ALBIDA, *al*-bid-a, white.
AUBRIETIOIDES, aw-bre-te-*oy*-des, like an Aubrieta.
BELLIDIFOLIA, bel-id-if-*o*-le-a, daisy-leaved.
LUCIDA, *lu*-sid-a, shining, *i.e.*, the leaves.
PETRÆA, pet-*re*-a, of rocks.

**Aralia,** ar-*a*-le-a; derivation uncertain, possibly from a vernacular name. Herbaceous plants and shrubs.

CASHMIRICA, kash-*mir*-ik-a, of Cashmere.

CHINENSIS, tshi-*nen*-sis, Chinese.

ELEGANTISSIMA, el-e-gan-*tis*-sim-a, most elegant.

JAPONICA, jap-*on*-ik-a, of Japan.

SIEBOLDTI, see-bold-e, after Siebold.

SPINOSA, spi-*no*-sa, spiny.

VEITCHI, *veech*-e after James Veitch.

**Araucaria,** a-raw-*kar*-e-a; from *araucanos*, name of the Indian tribe of Province of Araneo, Chile, where A. imbricata was first found. Coniferous trees.

EXCELSA, eks-*sel*-sa, lofty. The Norfolk Island Pine.

IMBRICATA, im-bre-*ka*-ta, overlapping, *i.e.*, the leaves. The Monkey Puzzle.

**Araujia,** a-*rau*-je-a; the Brazilian name. Tropical climber.

GRAVEOLENS, *grav*-e-ol-ens, strong-smelling.

**Arbutus,** *ar*-but-us; Latin name for A. Unedo, the Strawberry tree; some authorities derive word from Celtic, *arboise*, rough-fruited; commonly pronounced ar-*bu*-tus. Small trees.

MENZIESII, men-*ze*-sei or (Scots), ming-*es*-ei, after Menzies, the Scottish botanist.

UNEDO, u-*ned*-o or u-*ne*-do, meaning obscure. Pliny, the Roman naturalist, derives word from *unus*, one, and *edo*, to eat, *i.e.*, to eat one only—pleasant but unwholesome! The Strawberry Tree.

**Arctostaphylos,** ark-tos-*taf*-il-os; from Gr. *arktos*, a bear, and *staphyle*, a bunch of grapes, the berries of some species being eaten by bears, hence Bear-berry. Shrubs.

MANZANITA, man-zan-*it*-a, a Spanish-Californian name for the genus generally.

UVA-URSI, *u*-va-*ur*-se, bear's grape. The Bear-berry.

**Arctotis,** ark-*to*-tis; from Gr. *arktos*, a bear, and *ous*, an ear, probably in reference to shaggy fruit. Annuals and perennials.

ASPERA, *as*-per-a, rough, *i.e.*, the leaves.

BREVISCAPA, brev-is-*ka*-pa, short-stalked.

GRANDIS, *gran*-dis, splendid.

LEPTORHIZA, lep-to-*re*-za, having fine (slender) roots.

SCAPIGERA, skap-*ij*-er-a, bearing scapes.

**Ardisia,** ar-*dis*-e-a; from *ardis*, a spearhead, alluding to the shape of the anthers. Greenhouse berry-bearing shrub.

CRENULATA, kren-ul-*a*-ta, crenate or round-notched—the leaves.

**Areca,** ar-*e*-ka; from *areec*, a native name. Tropical palm. The Areca Nut.

LUTESCENS, lu-*tes*-senz, becoming yellow.

**Arenaria,** ar-en-*ar*-e-a; from L. *arena*, sand, *i.e.*, inhabiting sandy places, hence Sandwort. Rock plants.

BALEARICA, bal-e-*ar*-ik-a, of Balearic Islands.

GOTHICA, go-thik-a, of Gothland, N. Germany.

LARICIFOLIA, lar-is-if-*o*-le-a, Larix (larch)-leaved.

LEDEBOURIANA, led-e-boo-re-*a*-na, after Ledebour, a professor of botany.

MONTANA, mon-*ta*-na, of mountains.

PURPURESCENS, pur-pur-*es*-ens, purple—the flowers.

TETRAQUETRA, tet-ra-*kwet*-ra, four-angled—the leaves in fours.

VERNA, *ver*-na, spring—time of flowering.

VILLARSI, vil-*ar*-se, after Villars, a professor of botany.

**Arequipa,** ar-e-*quip*-a; the name of the town in Peru where the species is found. Greenhouse cacti.

LEUCOTRICHUS, loo-ko-*trik*-us, white-haired, having white hairs.

**Argemone,** ar-*gem*-o-ne; from Gr. *argemos*, a white spot (cataract) on the eye which the plant was supposed to cure; or from Gr. *argos*, slothful, *i.e.*, from the narcotic effects of the poppy. Annuals and perennials. The Mexican Poppy.

GRANDIFLORA, gran-dif-*lo*-ra, large-flowered.

MEXICANA, meks-e-*kan*-a, of Mexico.

PLATYCERAS, plat-y-*se*-ras, having broad prickles.

**Argyroderma,** ar-gy-rod-*er*-ma; from Gr. *argyros*, silver, and *derma*, skin, referring to the silvery deposit on the skin. Greenhouse succulent.

TESTICULARE, tes-tik-u-*lar*-e, tubercle-like.

**Ariocarpus,** *ar*-e-o-*kar*-pus; compound of *Aria* (the Whitebeam—Pyrus Aria) and Gr. *carpos*, fruit, the fruits suggesting those of the Whitebeam. Greenhouse cacti.

FISSURATUM, fis-sur-*a*-tum, cleft.

**Ariocarpus** (*continued*)
RETUSA, ret-*u*-sa, blunt.
PRISMATICUM, priz-*mat*-ik-um, cut like a prism.

**Arthropodium**, arth-ro-*pod*-e-um; from Gr. *arthron*, a joint, and *pous*, a foot; the floral foot stalks are jointed. Greenhouse herbaceous plants.
CIRRHATUM, kir-*ha*-tum, having curls.
PANICULATUM, pan-ik-ul-*a*-tum, bearing tufts or panicles.

**Arisæma**, ar-is-*e*-ma; Gr. name referring to red-blotched leaves of some species. Tuberous-rooted perennials.
SPECIOSUM, spes-e-*o*-sum, showy.
TRIPHYLLUM, trif-*il*-lum, three-leaved, *i.e.*, leaf divisions.
RINGENS, *rin*-gens, gaping—the open spathe.

**Arisarum**, ar-is-*ar*-um; possibly from Gr. *arista*, a bristle, or awn, and *arum*, to which the genus is allied, the spathe, or sheath of the flower, having a spike. Herbaceous perennials.
PROBOSCIDEUM, pro-bos-*sid*-e-um, having a proboscis or tail-like appendage; lit. like a snout.
VULGARE, vul-*gar*-e, common.

**Aristolochia**, ar-is-to-*lo*-ke-a; from Gr. *aristos*, best (most useful), and *locheia*, childbirth, alluding to ancient use in maternity—the Birthwort. Tropical and hardy climbing plants and shrubs.
CLEMATITIS, klem-at-*i*-tis, clematis-like.
ELEGANS, *el*-e-ganz, elegant.
GIGAS, *ji*-gas, of giant proportions—the flower.
SIPHO, *si*-fo, siphon or tube-bearing, hence Eng. name, Dutchman's pipe.

**Aristotelia**, ar-is-to-*te*-le-a; believed to be named in honour of the philosopher Aristotle. Shrubs.
MACQUI, *mak*-u-e, after Macqui, a Chilean.

**Armeria**, ar-*meer*-e-a; old L. name. English name, Thrift or Sea Pink. Rock and border plants.
CÆSPITOSA, kees-pit-*o*-sa (or ses-pit-*o*-sa), closely-tufted (lit. turf-like).
CEPHALOTES, sef-al-*o*-tes, large heads of flowers.
FASCICULATA, fas-sik-ul-*a*-ta, bundled, *i.e.*, the flowers.
JUNCEA, *jun*-ke-a, rush-like.
LATIFOLIA, lat-if-*o*-le-a, broad-leaved.

**Armeria** (*continued*)
MARITIMA, mar-*it*-im-a, of the sea. Sea Pink or Thrift.
PLANTAGINEA, plan-ta-*jin*-e-a, plantain (Plantago)-like—the leaves.
SPLENDENS, *splen*-denz, splendid.

**Arnebia**, ar-*ne*-be-a; from *arneb*, Arabic name for one of the species. Rock or border plant.
ECHIOIDES, ek-e-*oy*-des, like an Echium. The Prophet's Flower.

**Arnica**, *ar*-nik-a; origin uncertain, possibly from Gr. *arnakis*, a lamb's skin (or Gr. *arnion*, a lamb), the leaves having a soft texture. Herbaceous plants.
AMPLEXICAULIS, am-pleks-e-*kaw*-lis, with leaves clasping the stem.
MONTANA, mon-*ta*-na, of mountains.
SACHALINENSIS, sak-al-in-*en*-sis, of Sakhalin Island.

**Artemisia**, ar-tem-*ees*-e-a; called after Artemis (Diana), one of the divinities of ancient Greece. Perennials.
ABROTANUM, ab-*rot*-a-num, L. name for Southernwood.
ABSINTHIUM, ab-*sin*-the-um, L. name for Wormwood.
ALPINA, al-*pine*-a, Alpine.
ARBORESCENS, ar-bor-*es*-senz, tree-like.
BOREALIS, bor-e-*a*-lis, of the North.
CANA, *ka*-na, hoary.
DRACUNCULUS, drak-*un*-ku-lus, dragon-like. The Tarragon.
FRIGIDA, *frij*-id-a, cold, *i.e.*, frosty-looking.
GNAPHALIOIDES, naf-a-le-*oy*-des, resembling Gnaphalium (Cudweed).
GRACILIS, *gras*-il-is, slender.
LACTIFLORA, lak-tif-*lo*-ra, flowers milk-white.
LANATA, lan-*a*-ta, woolly.
LAXA, *laks*-a, loose, or open.
LUDOVICIANA, lu-do-*vis*-e-a-na, after Ludovic.
PEDEMONTANA, ped-e-mon-*ta*-na, from Piedmont.
PONTICA, *pon*-tik-a, Pontus, the shores of the Black Sea.
SCOPARIA, sko-*par*-e-a, having twiggy branches, like Cytisus scoparius.
STELLARIANA, stel-lar-e-*a*-na, starry.
TANACETIFOLIA, tan-a-set-if-*o*-le-a, leaves like Tanacetum (Tansy).

**Arum**, *a*-rum; ancient name, possibly from Arabic *ar*, fire, in reference to burning taste of plant. Herbaceous perennials.
DRACUNCULUS, drak-*un*-ku-lus, dragon-like. The Dragon Arum.
ITALICUM, it-*al*-ik-um, Italian.

12

**Arum** (*continued*)

MACULATUM, mak-ul-*a*-tum, spotted, *i.e.*, the leaves.

PALÆSTINUM, pal-es-*teen*-um, of Palestine.

**Arundinaria**, ar-un-din-*a*-re-a; from L. *arundo*, a reed. Bamboos.

ANCEPS, *an*-seps, two-edged, flattened.

AURICOMA, awr-*ik*-o-ma, golden-haired, *i.e.*, variegated with yellow stripes.

FORTUNEI, *for*-tu-nei, after Fortune, botanist and collector.

METAKE, met-*a*-ke, the Japanese name.

NITIDA, *nit*-id-a, shining, glossy.

PALMATA, pal-ma-ta, palmate or handleaved.

PUMILA, *pu*-mil-a, small, dwarf.

VEITCHII, *veech*-ei, after James Veitch.

**Arundo**, ar-*un*-do; from L. *arundo*, a reed. Reeds.

DONAX, *do*-naks, name for the Great Reed of Provence.

PHRAGMITES, frag-*my*-tez, Gr. name for Reed. Common Reed.

**Asarum**, as-*ar*-um; from ancient name meaning not clear. Herbaceous plants.

EUROPÆUM, u-ro-*pe*-um, of Europe.

VIRGINICUM, vir-*jin*-ik-um, of Virginia.

**Asclepias**, as-*kle*-pe-as; after Æsculapius, a Greek, who was learned in the medicinal properties of plants. Herbaceous and sub-shrubby.

CURASSAVICA, ku-ras-*sav*-ik-a, Curassivian.

INCARNATA, in-kar-*na*-ta, flesh-coloured.

OBTUSIFOLIA, ob-tu-sif-*o*-le-a, bluntleaved.

TUBEROSA, tu-ber-*o*-sa, tuberous.

**Asparagus**, as-*par*-ag-us; ancient Gr. name, said to be derived from Gr. *a*, intensive, and *sparasso*, to tear, alluding to the prickles of some species. Herbaceous and climbing plants.

OFFICINALIS, of-fis-in-*a*-lis, of shops. The Culinary Asparagus.

PLUMOSUS, plu-*mo*-sus, plumed or feathery. The so-called "Asparagus Fern."

SPRENGERI, *spreng*-er-e, after Sprenger, professor of botany.

**Asperula**, as-*per*-u-la; from L. *asper*, rough, alluding to the leaves. Rock plants.

ARCADIENSIS, ar-ka-de-*en*-sis, Arcadian.

AZUREA, az-*u*-re-a, pale blue.

CYNANCHICA, si-*nan*-chik-a, Gr. name for quinsy, The Squinancy-wort.

GUSSONII, *gus*-on-ei, of Gussone, a professor of botany.

**Asperula** (*continued*)

HEXAPHYLLA, heks-af-*il*-la, having six leaves or leaflets.

HIRTA, *hir*-ta, rough or shaggy.

ODORATA, od-o-*ra*-ta, sweet-smelling. The Sweet Woodruff.

ORIENTALIS, or-e-en-*ta*-lis, Eastern.

SUBEROSA, su-ber-*o*-sa, corky, *i.e.*, the stems.

**Asphodeline**, as-fod-el-*e*-ne; allied to Asphodel. Herbaceous plants.

LIBURNICA, li-*ber*-nik-a, of Liburnia, Eastern Adriatic.

LUTEA, lu-*te*-a, yellow.

TENUIOR, ten-*u*-e-or, more slender.

**Asphodelus**, as-*fod*-el-us; probable derivation, Gr. *a*, not, and *sphallo*, to supplant, the stately flowers not being easily surpassed. Herbaceous plants.

FISTULOSUS, fis-tu-*lo*-sus, pipe-like, hollow.

RAMOSUS, ra-*mo*-sus, with many branches. The Asphodel.

**Aspidistra**, as-pe-*dis*-tra; from Gr. *aspidion*, a little shield, referring to the form of the flower, or perhaps the mushroom-shaped stigma. Greenhouse or room plants.

ELATIOR, e-*la*-te-or, taller.

LURIDA, *lu*-rid-a, sallow-coloured—the flowers.

**Aspidium**, as-*pid*-e-um; from Gr. *aspidion*, a little shield, alluding to the shape of the spore-covering or indusium. Greenhouse and hardy ferns.

ACROSTICHOIDES, a-kros-tik-*oy*-des, like acrosticum.

ACULEATUM, a-ku-le-*a*-tum, prickly. The Shield Fern.

ANGULARE, ang-ul-*a*-re, angular. The Soft Shield Fern.

CAPENSE, ka-*pen*-se, of the Cape.

FALCATUM, fal-*ka*-tum, hooked—curved pinnæ. The False Holly Fern.

LONCHITIS, lon-*ki*-tis, spear-shaped. The Holly Fern.

MUNITUM, mu-*ne*-tum, armed with spines.

VESTITUM, ves-*tee*-tum, clothed.

**Asplenium**, as-*ple*-ne-um; from Gr. *a*, not, and *splene*, spleen, the Black Spleenwort (A. adiantum nigrum) once being regarded as a cure for diseases of the spleen. Greenhouse and hardy ferns.

ADIANTUM-NIGRUM, ad-i-*an*-tum, the Maidenhair, *ni*-grum, black, the stems, the Black Maidenhair Spleenwort.

BULBIFERUM, bul-*bif*-er-um, bulbil-bearing.

## Asplenium (continued)

CETERACH, set-er-ak, origin obscure; said to be a Persian name. The Scale Fern.

EBENEUM, eb-e-ne-um, ebony-stalked.

FILIX-FŒMINA, fe-liks-fem-in-a, female fern. The Lady Fern—in allusion to its elegance.

FONTANUM, fon-ta-num, growing by a fountain or spring.

GERMANICUM, jer-man-e-kum, of Germany.

LUCIDUM, lu-sid-um, shining.

NIDUS, ny-dus, L. for nest. Bird's-nest Fern.

MARINUM, mar-e-num, sea. The Sea Spleenwort.

MONANTHEMUM, mon-anth-em-um, one flowered, i.e., one sorus per pinna.

RUTA-MURARIA, roo-ta-mu-ra-re-a, rue of the wall. The Wall-rue Fern.

THELYPTEROIDES, thel-ip-ter-oy-des, like a thelypteris.

TRICHOMANES, trik-om-an-ez, a thin hair or bristle. The Maidenhair Spleenwort.

VIRIDE, vir-id-e, green.

VIVIPARUM, vi-vip-ar-um, plant-bearing —plantlets on the fronds.

**Aster,** as-ter; from Gr. aster, a star, which the flower is supposed to resemble. Herbaceous perennials. Michaelmas Daisy. Many hybrid forms.

ACRIS, ak-ris, acrid, pungent.

ALPINUS, al-pine-us, alpine.

AMELLUS, a-mel-lus, name given by Virgil to a blue aster-like plant by the River Mella.

CORDIFOLIA, kor-dif-o-le-a, heart-shaped leaves.

DELAVAYI, del-a-va-i, of Abbé Delavay, a missionary.

DIFFUSUS, dif-fu-sus, spread out.

DUMOSUS, du-mo-sus, bushy.

ERICOIDES, er-ik-oy-des, like Erica (heather).

FARRERI, far-rer-i, of Farrer.

LÆVIS, le-vis, small, polished.

LINOSYRIS, lin-os-er-is, flax (Linum)-like. The Goldilocks.

NOVÆ-ANGLIÆ, no-ve-ang-le-e, of New England, U.S.A.

NOVI-BELGII, no-vi-bel-jei, of New York, name of historical origin.

PUNICEUS, pu-nis-e-us, purple.

SUBCŒRULEUS, sub-se-ru-le-us, somewhat or slightly blue.

VIMINEUS, vim-in-e-us, with long, pliant growths, like an Osier.

YUNNANENSIS, yun-nan-en-sis, of Yunnan, China.

**Astilbe,** as-til-be; from a, no, and stilbe, brightness, meaning obscure; possibly

## Astilbe (continued)

refers to many of older species which had colourless flowers. Herbaceous plants.

ARENDSII, ar-ends-ei, after Arends.

CHINENSIS, tshi-nen-sis, Chinese.

DAVIDII, da-vid-ei, of Abbé David, a missionary.

JAPONICA, jap-on-ik-a, of Japan.

RIVULARIS, riv-u-lar-is, of streams.

SIMPLICIFOLIA, sim-plis-if-o-le-a, having simple (not compound) leaves.

THUNBERGII, thun-ber-gei, after Thunberg, a botanist.

**Astragalus,** as-trag-a-lus; from Gr. astragalos, one of the bones of the human ankle-bone, alluding to shape of the seed; another explanation is suggested by supposed likeness of the root to the ankle-bone. Herbaceous plants and shrubs.

ALOPECUROIDES, al-o-pek-u-roy-dez, foxtail-like.

MONSPESSULANUS, mon-spes-sul-a-nus, of Montpellier.

ONOBRYCHIS, on-o-bri-kis, lit. an ass's fodder. The Sainfoin.

TRAGACANTHA, trag-a-kan-tha, goat'sthorn (from Gr. tragos, goat and akantha, thorn).

**Astrantia,** as-tran-te-a; from Gr. aster, a star, referring to the star-like flower umbels. Herbaceous plants.

BIEBERSTEINII, bi-ber-sti-nei, after Bieberstein, Russian botanist.

CARNIOLICA, kar-ne-ol-ik-a, Carniolian.

MAJOR, ma-jor, greater.

MINOR, mi-nor (or myn-or), smaller.

**Atragene,** at-rag-en-e; believed to be from Gr. athro, pressure, and gennao, to produce, referring to the clinging tendrils; name first applied to a species of clematis and now included in that genus. Climbing shrubs.

ALPINA, al-pine-a, alpine.

**Atriplex,** a-trip-leks; ancient L. name, many suggested derivations, perhaps most popular being Gr. a, no, traphein, nourishment, several species growing in arid soils of deserts. Shrubs.

HALIMUS, hal-im-us, old generic name.

HORTENSIS, hor-ten-sis, of gardens. The Orache.

**Astrophytum,** as-tro-fy-tum; from Gr. aster, a star, and phytos, a plant, the plants are star-shaped. Greenhouse cacti.

MYRIOSTIGMA, my-re-os-tig-ma, manydotted.

**Atropa**, *at*-ro-pa; from Gr. *atropos*, one of the Three Fates of Grecian mythology from whom there was no escape, alluding to the poisonous berries. Herbaceous.

> BELLADONNA, bel-la-*don*-na, old generic name. The Belladonna or Deadly Nightshade.

**Aubrieta**, aw-*bre*-she-a (or o-*bre*-she-a); from M. Aubriet, a French botanical artist. Rock plants.

> DELTOIDEA, del-*toy*-de-a, three-angled, like the Greek letter *Δ* (delta), and *oides*, like, said to allude to triangular petals. Many florists' forms.

**Aucuba**, aw-*ku*-ba (or *aw*-kub-a); the Japanese name. Evergreen shrubs.

> JAPONICA, jap-*on*-ik-a, of Japan. Variety maculata (spotted) is the familiar variegated kind.

**Avena**, a-*ve*-na; from L. *avena*. A grass.

> STERILIS, ster-*il*-is, barren.

**Azalea**, az-*a*-le-a; Gr. *azaleos*, dry or parched, probably from A. pontica, inhabiting dry situations. All species now under Rhododendron. Flowering shrubs, greenhouse and hardy.

> INDICA, *in*-dik-a, Indian—the familiar evergreen greenhouse azalea in many hybrid varieties.
>
> KURUME, *ku*-rume, Japanese name. Hybrid forms of A. (Rhododendron) amœna.
>
> MOLLIS, *mol*-lis, soft—the leaves are downy.

**Azara**, az-*ar*-a; named after J. N. Azara, a Spanish patron of botany. Shrubs.

> DENTATA, den-*ta*-ta, toothed, the leaves.
>
> GILLIESII, *gil*-lies-ei, after Gillies.
>
> MICROPHYLLA, mi-kro-*fil*-la, small-leaved.

**Azolla**, a-*zol*-la; from Gr. *azo*, to dry and *ollo*, to kill—killed by dryness. Greenhouse floating aquatic.

> CAROLINIANA, kar-o-*lin*-e-an-a, from Carolina.

**Babiana**, bab-e-*a*-na; from *babianer*, Dutch name for baboon, which is said to devour the bulbs, hence Baboon-root. Tender bulbous plants.

> DISTICHA, *dis*-tik-a, in two rows.
>
> PLICATA, plik-*a*-ta, folded.
>
> STRICTA, *strik*-ta, upright.

**Baccharis**, *bak*-a-ris; from Gr. *Bacchus*, the god of wine, a spicy extract from some species having been used for mixing with wine. Shrubs.

> HALIMIFOLIA, hal-im-if-*o*-le-a, leaves like Atriplex Halimus. The Tree Ground-sel.

**Bambusa**, bam-*bu*-sa; aboriginal name in Malaya. See Arundinaria and other genera.

> PALMATA, pal-*ma*-ta, hand-shaped—the arrangement of the leaves.

**Banksia**, *bangk*-se-a; after Sir Joseph Banks, famous British scientist. Greenhouse shrubs.

> DRYANDROIDES, dry-an-*droy*-dez, resembling Dryandra.
>
> QUERCIFOLIA, kwer-ke-*fol*-e-a, oak (Quercus)-leaved.

**Baptisia**, bap-*tis*-e-a; from Gr. *bapto*, to dye, some species yielding dyers' tinctures. Hardy herbaceous plants.

> AUSTRALIS, aws-*tra*-lis, southern.
>
> TINCTORIA, ting-*tor*-e-a, of dyers.

**Barbarea**, bar-bar-*e*-a; called after St. Barbara, to whom the Winter Cress was dedicated. Herbaceous salad plant. Winter or American Cress.

> PRÆCOX, *pre*-koks, early.
>
> VULGARIS FLORE PLENO, vul-*gar*-is *flor*-e *plen*-o, double-flowered common. The Double Yellow Rocket.

**Bartonia**, bar-*to*-ne-a; after Dr. Barton, botanist of Philadelphia. Half-hardy annual.

> AUREA, *aw*-re-a, golden yellow.

**Bauhinia**, baw-*in*-i-a; commemorates John and Caspar Bauhin, botanists. Warm-house evergreen flowering shrubs.

> GRANDIFLORA, gran-dif-*lor*-a, large-flowered.
>
> NATALENSIS, na-tal-*en*-sis, from Natal.
>
> PURPUREA, pur-*pur*-e-a, purple.

**Beaucarnea**, bo-*kar*-ne-a; probably commemorative, but of whom or why are unknown. Greenhouse ornamental leaved herbaceous plants.

> GLAUCA, *glaw*-ka, sea-blue or glaucous—the leaves.
>
> RECURVATA, rek-ur-*va*-ta, recurved—the leaves.

**Begonia**, be-*go*-ne-a; after Michael Begon, a French botanist and patron. Most varieties in general culture are of garden origin. Tender bedding and indoor perennials; flowering and ornamental foliage.

> ALBO-COCCINEA, *al*-bo-kok-*sin*-e-a, white and scarlet.
>
> BOLIVIENSIS, bol-iv-e-*en*-sis, Bolivian
>
> COCCINEA, kok-*sin*-e-a, scarlet.
>
> EVANSIANA, ev-*an*-se-*a*-na, after Evans.

**Begonia** (*continued*)

FRŒBELII, fre-*bel*-ei, after Frœbel, a botanist.

FUCHSIOIDES, few-she-*oy*-dez, fuchsia-like.

GRACILIS, *gras*-il-is, graceful.

HAAGEANA, *haag*-e-an-a, after Haage.

HYDROCOTYLIFOLIA, hy-dro-*kot*-il-if-*o*-le-a, Hydrocotyle (Pennywort)-leaved.

INCARNATA, in-kar-*na*-ta, flesh-coloured.

LLOYDII, *loy*-dei, after Lloyd.

MACULATA, mak-ul-*a*-ta, blotched—the leaves.

MANICATA, man-ik-*a*-ta. collared (sleeved) with fleshy, scale-like bristles on the leaf-stalk.

METALLICA, me-*tal*-ik-a, metallic—the sheen of the foliage.

OCTOPETALA, ok-to-*pet*-a-la, eight-petal-led.

OLBIA, olb-*e*-a, rich.

REX, reks, a king, presumably the handsome foliage.

SCANDENS, *skan*-dens, climbing.

SOCOTRANA, sok-*o*-tra-na, of Socotra.

SEMPERFLORENS, sem-per-*flo*-rens, always flowering.

TUBEROSA, tew-ber-*o*-sa, tuberous.

WELTONIENSIS, wel-ton-e-*en*-sis, of Welton.

WORTHIANA, wurth-e-*a*-na, after Worth.

**Bellidiastrum,** bel-lid-e-*as*-trum; from L. *bellis*, a Daisy, and *astrum*, a star. Rock plant.

MICHELII, *mi*-kel-ei, of Micheli, Italian botanist.

**Bellis,** *bel*-lis; from L. *bellus*, pretty; Ang.-Sax. *daeges eage*, day's-eye, the Daisy. Herbaceous and rock plants.

PERENNIS, per-*en*-nis, perennial. The Daisy.

ROTUNDIFOLIA, ro-tun-dif-*o*-le-a, round-leaved.

SYLVESTRIS, sil-*ves*-tris, of woods.

**Bellium,** *bel*-le-um; derivation as above, the flowers being like a Daisy. Rock plants.

BELLIDIOIDES, bel-lid-e-*oy*-dez, daisy-like. The False Daisy.

MINUTUM, min-*u*-tum, small, minute.

**Beloperone,** bel-o-*per*-o-ne, from Gr. *belos*, an arrow and *peronne*, a hand. The form of the connectivum. Warm-house flowering plants.

VIOLACEA, vi-o-*la*-ce-a, violet coloured.

**Benthamia,** ben-*tha*-me-a; after George Bentham, the English botanist. Now included under Cornus. Shrubs.

FRŒGIFERA, frag-*if*-er-a, strawberry-like —the fruits.

**Berberidopsis,** ber-ber-e-*dop*-sis; from *Berberis*, and Gr. *opsis*, a resemblance, being like a Berberis. Twining shrub.

CORALLINA, kor-a-*line*-a, coral—the flowers.

**Berberis,** *ber*-ber-is; from the Arabic name *berberys*. Shrubs.

AGGREGATA, ag-gre-*ga*-ta, heaped to-gether—the clustered fruits.

AQUIFOLIUM, ak-we-*fo*-le-um, sharp, or holly-leaved.

ARISTATA, ar-is-*ta*-ta, bristled.

BUXIFOLIA, buks-if-*o*-le-a, Buxus (box)-leaved.

DARWINII, *dar*-win-ei, after Charles Darwin who discovered it in 1835 when attached to the "Beagle."

DICTYOPHYLLA, dik-ti-*of*-il-la, with net-veined leaves.

EMPETRIFOLIA, em-pet-rif-*o*-le-a, Em-petrum-leaved.

GAGNEPAINII, *gag*-ne-pain-ei, after Gag-nepain.

HOOKERI, *hoo*-ker-i, after W. J. Hooker, the botanist.

ILICIFOLIA, il-lis-if-*o*-le-a, holly-leaved.

JAPONICA, jap-*on*-ik-a, Japanese.

LYCIUM, *lis*-e-um, like the genus Lycium.

NEPALENSIS, nep-al-*en*-sis, of Nepal.

NERVOSA, ner-*vo*-sa, large-nerved.

POLYANTHA, pol-e-*an*-tha, many-flow-ered.

PRUINOSA, pru-in-*o*-sa, frosted, *i.e.*, the glaucous fruits.

REPENS, *re*-pens, creeping.

SANGUINEA, san-*gwin*-e-a, blood-red—the calyx.

SARGENTIANA, sar-jen-te-*a*-na, after Prof. Sargent.

SINENSIS, sin-*en*-sis, of China.

STENOPHYLLA, sten-o-*fil*-la, narrow-leaved.

THUNBERGII, thun-*berj*-ei, after Thun-berg, botanist.

VERRUCULOSA, ver-ru-ku-*lo*-sa, warted—the bark.

VIRESCENS, vir-*es*-sens, greenish—the flowers.

VULGARIS, vul-*gar*-is, common. The Barberry.

WILSONÆ, *wil*-son-e, after Mrs. E. H. WILSON.

**Berchemia,** ber-*she*-me-a; origin uncer-tain. Climbing shrubs.

FLAVESCENS, fla-*ves*-sens, yellowish.

VOLUBILIS, vol-*u*-bil-is, twining.

**Bergerocactus,** ber-ger-o-*kak*-tus; after Alwin Berger, curator of Hanbury Gar-den at La Mortola, Italy. Greenhouse cacti.

EMORYI, e-*mor*-e-i, after Emory.

**Bertolonia,** ber-tol-*o*-ne-a; after A. Bertoloni, an Italian botanist. Dwarf tropical plants with ornamental foliage.
GUTTATA, gut-*ta*-ta, speckled or spotted.
MACULATA, mac-ul-*a*-ta, spotted.
MARMORATA, mar-mor-*a*-ta, like marble.
VAN HOUTTEI, van-*hout*-te-i, after van Houtte.

**Bessera,** *bes*-ser-a; after Dr. Besser, professor of botany. Greenhouse bulbous plant.
ELEGANS, *el*-e-ganz, elegant.

**Beta,** *be*-ta; from L. *beta,* Beetroot, or Celtic *bett,* red. Biennials. Ornamental leaved and culinary.
CICLA, *sik*-la, old name (Sicilian). Silver Beet.
C. VARIEGATA, var-e-eg-*a*-ta, vari-coloured. Chilean Beet.
VULGARIS, vul-*gar*-is, common. The Beetroot.

**Betonica,** bet-*on*-ik-a; from Celtic *bentonic—ben,* head, and *ton,* good, referring to herbalists' use of common kind. Herbaceous plants.
GRANDIFLORA, gran-dif-*lo*-ra, large-flowered.

**Betula,** *bet*-u-la; the Latin name for Birch; Ang.-Sax. *birce.* Trees.
ALBA, *al*-ba, white, the stem.
HUMILIS, *hum*-il-is, lowly—on the ground.
NANA, *na*-na, dwarf.
NIGRA, *ni*-gra, black—the bark.
PAPYRIFERA, pa-pir-*if*-er-a, papery—the bark.
PENDULA, *pen*-du-la, pendulous or "weeping."
POPULIFOLIA, pop-u-lif-*o*-le-a, poplar (Populus)-leaved.
VERRUCOSA, ver-ru-*ko*-sa, warted—the young wood.

**Bidens,** *bi*-dens; from L. *bi,* two, and *dens,* teeth, the seed having two toothlike projections. Half-hardy herbaceous.
DAHLIOIDES, day-le-*oy*-dez, dahlia-like.

**Bignonia,** big-*no*-ne-a (or be-*no*-ne-a); after Abbé Bignon, librarian to Louis XIV. Greenhouse and hardy climbers.
CAPREOLATA, kap-re-o-*la*-ta, tendrilled.
TWEEDIANA, *tweed*-e-a-na, after Tweedie.
VENUSTA, ven-*us*-ta, lovely.

**Billardiera,** bil-lard-e-*air*-a; named after the French botanist Labillardière. Climbing shrub.
LONGIFLORA, long-if-*lo*-ra, long-flowered.

**Billbergia,** bil-*ber*-je-a; after J. G. Billberg, a Swedish botanist. Tropical herbaceous, flowering plants.
LIBONIANA, lib-o-ne-*a*-na, of Libon, Brazil.
MARMORATA, mar-mor-*a*-ta, marbled.
NUTANS, *nu*-tanz, nodding.
SAUNDERSII, *saun*-derz-ei, after Saunders.
VITTATA, vit-*ta*-ta, striped.
ZEBRINA, ze-*bry*-na, zebra-striped.

**Blechnum,** *blek*-num; from *blechnon,* a Greek name for a fern. Hardy and greenhouse ferns.
BOREALE, bor-e-*a*-le, northern.
BRAZILIENSE, braz-il-e-*en*-se, Brazilian.
OCCIDENTALE, ok-se-den-*ta*-le, western.
SPICANT, *spe*-kant, spiked—the appearance of the fertile fronds.

**Bletia,** *blet*-e-a; after Don Louis Blet, a Spanish botanist. Terrestrial orchids.
HYACINTHINA, hy-a-sinth-*e*-na, hyacinth-like.

**Blitum,** *bly*-tum; from Celtic *blith,* insipid, referring to the fruits. Hardy annual.
CAPITATUM, kap-it-*a*-tum, in heads—the fruits. Strawberry Spinach or Strawberry Blite.

**Blumenbachia,** blu-men-*bak*-e-a; after John Frederick Blumenbach, M.D., of Göttingen. Tender annuals.
LATERITIA, lat-er-*it*-e-a, brick-red—the colour of the flowers.

**Bocconia,** bok-*ko*-ne-a; after Boccone, an Italian botanist. Herbaceous plants.
CORDATA, kor-*da*-ta, heart-shaped—the leaves.
MICROCARPA, mi-kro-*kar*-pa, bearing small fruits.

**Boltonia,** bol-*to*-ne-a; after J. B. Bolton, professor of botany. Herbaceous plants.
ASTEROIDES, as-ter-*oy*-des, aster-like.

**Bomarea,** bo-*ma*-re-a; after French botanist. Greenhouse flowering twiners.
CARDERI, kar-*der*-i, after Carder.
CONFERTA, kon-*fer*-ta, dense flowered.
OLIGANTHA, ol-ig-*an*-tha, few flowered.
PATOCOCENSIS, pat-ok-o-*ken*-sis, of Patococha, Ecuador.

**Borago,** bor-*a*-go; derivation uncertain, may be from mediæval L. *borra* or *burra,* rough hair in reference to rough foliage. Linnæus states name to be corruption of *corago* (L. *cor,* the heart, *ago,* to act) from its use in medicine as a heart sedative. Hardy annual and biennial herbs.

**Borago** (*continued*)

LAXIFLORA, laks-if-*lo*-ra, flowers in a loose spike.

OFFICINALIS, of-fis-in-*a*-lis, of the shop (herbal). The Borage.

**Boronia**, bor-*on*-e-a; after Francis Boroni, an Italian plant collector. Greenhouse flowering shrubby plants.

DRUMMONDII, drum-*mon*-dei, after Drummond.

ELATIOR, e-*la*-te-or, taller.

HETEROPHYLLA, het-er-of-*il*-la, leaves of varied shape.

MEGASTIGMA, meg-as-*tig*-ma, having a large stigma.

SERRULATA, ser-rul-*a*-ta, leaves finely toothed.

**Bougainvillea**, boo-gain-*vil*-le-a; after de Bougainville, a French navigator. Greenhouse shrubby climbers.

GLABRA, *gla*-bra, smooth—without hairs.

SPECIOSA, spes-e-*o*-sa, showy.

SPECTABILIS, spek-*tab*-il-is, notable.

**Boussingaultia**, boos-sin-*gault*-e-a; after Boussingault, a chemist of note.

BASELLOIDES, bas-el-*oy*-dez, Basella-like.

**Bouvardia**, boo-*vard*-e-a; after Dr. Chas. Bouvard of Paris. Greenhouse evergreen flowering plants of shrubby nature.

HUMBOLDTII CORYMBIFLORA, hum-*bolt*-i kor-im-bif-*lo*-ra, Humboldt's corymbflowered.

LONGIFLORA, long-if-*lo*-ra, long-flowered.

JASMINIFLORA, jas-min-if-*lo*-ra, jasmineflowered.

**Brachycome**, brak-e-*ko*-me; from Gr. *brachys*, short, *comus*, hair. Annuals.

IBERIDIFOLIA, i-ber-id-if-*o*-le-a, Iberis (candytuft)-leaved.

**Brassavola**, bras-sa-*vo*-la; after A. M. Brassavola, Italian botanist. Warm house orchids.

DIGBYANA, dig-by-*an*-a, after Digby.

**Brassia**, *bras*-se-a; after William Brass, plant collector. Warm-house orchids.

BRACHIATA, brak-e-*a*-ta, having arm-like divisions.

MACULATA, mak-ul-*a*-ta, spotted.

VERRUCOSA, ver-ru-*ko*-sa, warted.

**Brassica**, *bras*-sik-a; from L. *brassica* used by Pliny, from Celtic *bresic*, the name for cabbage. Culinary vegetables.

OLERACEA, ol-er-*a*-cea, as a herb. The Wild Cabbage.

**Brassica** (*continued*)

O. CAPITATA, kap-it-*a*-ta, having a head. The Garden Cabbage.

O. ACEPHALA, a-*kef*-a-la, or a-*sef*-a-la, without a head. The Borecole or Kale.

O. BOTRYTIS, bot-*ry*-tis, like a bunch of grapes. Broccoli. Cauliflower.

O. CAULIFLORA, kaul-if-*lo*-ra, stem flowered.

O. BULLATA, bul-*la*-ta, bubbles or blisters—on leaves. The Brussels Sprout.

O. B. MAJOR, *ma*-jor, larger. The Savoy.

O. CAULO-RAPA, *kaul*-o-ra-pa, turnipstemmed. The Kohl Rabi or Turnip Cabbage.

RAPA, *ra*-pa, of rape. The Turnip.

**Bravoa**, bra-*vo*-a; from Bravo, a botanist of Mexico. Bulbous plants.

GEMINIFLORA, jem-in-if-*lo*-ra, twin-flowered.

**Briza**, *bri*-za; from Gr. *brizo*, to nod. from the movement of the sprays, Grasses. Quaking grass.

MAXIMA, *maks*-im-a, greatest.

MEDIA, *me*-de-a, midway, medium.

**Brodiaea**, bro-*de*-e-a; named after J. J. Brodie, a Scottish botanist. Bulbous plants.

BRIDGESII, brid-*jees*-ei, after Bridges, a botanist.

CAPITATA, kap-it-*a*-ta, flowers clustered in a head.

CONGESTA, kon-*jes*-ta, crowded, the flowers.

GRANDIFLORA, gran-dif-*lo*-ra, largeflowered.

HENDERSONII, hen-der-*so*-nei, after Henderson.

HOWELLII, how-*el*-lei, after Howell.

IXIOIDES, iks-e-*oy*-dez, ixia-like.

LAXA, *laks*-a, loose—the flowers.

UNIFLORA, u-nif-*lo*-ra, one-flowered.

**Bromus**, *bro*-mus; from Gr. *broma*, fodder (*bromos*, a Wild Oat). Grasses.

BRIZÆFORMIS, bri-ze-*for*-mis, formed like Briza or Quaking Grass.

**Browallia**, brow-*al*-le-a; said to be named after J. Browallius, Bishop of Abo, Sweden. Greenhouse annuals.

ELATA, e-*la*-ta, tall, stately.

SPECIOSA, spes-i-*o*-sa, showy.

**Brunella**, broo-*nel*-la; a synonym of Prunella, which see.

**Brunfelsia**, brun-*felz*-e-a; after Otto Brunfels, a physician. Warm-house evergreen flowering shrubs.

CALYCINA, kal-ik-*een*-a, cup-shaped.

**Bryanthus,** bry-*an*-thus; from Gr. *bryon*, moss and *anthos*, flower, growing among mosses. Low shrubs.

EMPETRIFORMIS, em-pet-rif-*or*-mis, Empetrum-like.

**Bryophyllum,** bry-of-*il*-um; from Gr. *bryo*, to sprout, and *phyllon*, a leaf, alluding to the leaves bearing plantlets round their edges. Greenhouse succulents.

CALYCINUM, kal-is-*een*-um, cup-shaped.

**Buddleia,** *bud*-le-a; called after Rev. Adam Buddle, one-time vicar of Farnbridge, Essex. Tender and hardy shrubs.

ALTERNIFOLIA, alt-er-nif-*o*-le-a, leaves alternate.

COLVILEI, *col*-vil-*le*-i, after Colvile.

FALLOWIANA, fal-low-e-*a*-na, after Sergt. Fallow of Edinburgh.

GLOBOSA, glo-*bo*-sa, globular, the flower clusters.

NANHOENSIS, nan-ho-*en*-sis, of Nan-ho, China.

NIVEA, *niv*-e-a, snowy—leaves and shoots white.

VARIABILIS, var-e-*ab*-il-is, variable.

**Bulbocodium,** bul-bo-*ko*-de-dum; from L. *bulbus*, a globular root (bulb), and *kodion*, wool, with which the bulbs are covered. Bulbs.

VERNUM, *ver*-num, spring—the flowering period.

**Buphthalmum,** bup-*thal*-mum or buf-*thal*-mum; from Gr. *bous*, an ox, and *ophthalmos*, an eye, the large-rayed flower supposed to resemble the eye of an ox. Hardy herbaceous.

SALICIFOLIUM, sal-is-if-*o*-le-um, Salix (willow)-leaved.

SPECIOSUM, spes-e-*o*-sum, showy. The Ox-eye.

**Bupleurum,** bu-*plu*-rum; ancient Gr. name for an umbelliferous plant. Hardy shrubs.

FRUTICOSUM, .frut-ik-*o*-sum, shrubby.

**Burchellia,** bur-*chel*-le-a; after W. Burchell, plant collector. Warm-house evergreen shrub.

CAPENSIS, ka-*pen*-sis, of the Cape of Good Hope.

**Butomus,** *bu*-to-mus; from Gr. *bous*, an ox, and *temno*, to cut, the sharp-edged leaves (or acrid juice) being said to injure the mouths of cattle. Aquatic.

UMBELLATUS, um-bel-*la*-tus, umbelled—the flowers.

**Buxus,** *buks*-us; ancient Latin name. Box tree. Shrubs and trees.

MICROPHYLLA, mi-krof-*il*-a, small leaved.

SEMPERVIRENS, sem-per-*veer*-enz, evergreen. The Box Tree.

S. SUFFRUTICOSA, suf-frut-ik-*o*-sa, having a woody base. The Edging Box.

**Cabomba,** cab-*om*-ba; native Guiana name. Greenhouse submerged aquatic.

CAROLINIANA, car-ol-in-i-*a*-na, native of Carolina State.

ROSÆFOLIA, ro-ze-*fol*-e-a, a rosy-red-leaved variety of caroliniana.

**Cacalia,** ka-*ka*-le-a; ancient Gr. name, possibly from Gr. *kakos*, pernicious, and *lian*, very much, supposed to be harmful to the soil. Hardy annual.

COCCINIA, kok-*sin*-e-a, scarlet. The Tassel Flower.

**Cactus,** kak-tus; from Gr. *kaktos*, a name used by Theophrastus for an unknown prickly plant, now applied to a group of cacti of which the Melon cactus is the type and was the first cactus brought to Europe (1581). This plant was called Cactus melocactus, since changed to Melocactus communis. Cactus has now been dropped as a generic term and is the English name for members of the family Cactaceæ.

**Cæsalpinia,** seez-al-*pin*-e-a; after Andreas Cæsalpini, Italian botanist.

GILLIESII, gil-*les*-ei, after Gillies.

JAPONICA, jap-*on*-ik-a, of Japan.

SEPIARIA, se-pe-*a*-re-a, of hedges.

**Caladium,** kal-*a*-de-um; origin uncertain, said to be from *kale*, the native name of the tuberous root. Tropical tuberous-rooted foliage plants; many hybrids.

ARGYRITES, ar-ger-*ee*-tez, silvery, alluding to white variegation.

BICOLOR, *bik*-ol-or, two-coloured.

CANDIDUM, *kan*-did-um, white.

MINUS ERUBESCENS, *mi*-nus er-u-*bes*-senz, dwarf and becoming red.

**Calamintha,** kal-a-*min*-tha; from Gr. *kala*, good, and *mintha*, mint. Aromatic herbs.

ALPINA, al-*pine*-a, alpine.

GRANDIFLORA, gran-dif-*lo*-ra, large-flowered.

CAL-CAL

**Calandrinia,** kal-an-*drin*-e-a; after J. L. Calandrini, a Swiss botanist. Rock garden annuals and perennials.

GRANDIFLORA, gran-dif-*lo*-ra, large flowered.

UMBELLATA, um-bel-*la*-ta, umbelled.

**Calanthe,** kal-*an*-thee; from Gr. *kalos* beautiful, and *anthos*, a flower. Deciduous terrestrial orchids.

VEITCHII, *veech*-ei, after Veitch.

VESTITA, ves-*tee*-ta, clothed.

**Calathea,** kal-*ath*-e-a; from Gr. *kalathos*, a basket, referring to a native use of the tough fibrous leaves. Tropical ornamental leaved plants.

ILLUSTRIS, il-*lus*-tris, brilliant.

MASSANGEANA, mas-san-ge-*a*-na, after Massange.

ORNATA, or-*na*-ta, adorned.

REGALIS, re-*ga*-lis, royal, stately.

ROSEA-PICTA, *ro*-ze-a-*pik*-ta, rose-coloured.

SANDERIANA, san-der-e-*a*-na, after Sander, nurseryman.

VEITCHII, *veech*-ei, after Veitch, nurseryman.

ZEBRINA, ze-*bry*-na, zebra-striped.

**Calceolaria,** kal-se-o-*lair*-e-a; from L. *calceolus*, a slipper or little shoe, alluding to the shape of the flower. Half-hardy rock, herbaceous, and shrubs.

BIFLORA, bif-*lo*-ra, two-flowered.

BURBIDGEI, bur-*bij*-ei, after F. W. Burbidge, who raised it.

CORYMBOSA, kor-im-*bo*-sa, corymbose.

INTEGRIFOLIA, in-teg-rif-*o*-le-a, lit. whole leaves, *i.e.*, not broken at the edges.

RUGOSA, roo-*go*-sa, wrinkled.

VIOLACEA, vi-o-*la*-se-a, violet-like—the colour of the flowers.

**Calendula,** kal-*en*-du-la; from L. *calendae*, the first day of the month, probably alluding to the flowering of the plant throughout the year. Hardy annual.

OFFICINALIS, of-fis-in-*a*-lis, of herbal shops. The Pot Marigold.

**Calla,** *kal*-la; ancient name of unknown meaning. Aquatic.

PALUSTRIS, pal-*us*-tris, pertaining to marshes.

**Callirhoe,** kal-lir-*ho*-e; after *Callirhoe*, the name of a divinity of the ancient Greeks. Hardy herbaceous.

INVOLUCRATA, in-vol-u-*kra*-ta, the leaf-edges rolled together.

**Callistemon,** kal-lis-*tee*-mon; from Gr. *kallistos*, most beautiful, and *stemon*, a stamen, the beauty of the flowers residing in the coloured stamens. Greenhouse flowering shrubs.

LINEARIS, lin-e-*a*-ris, linear.

SPECIOSUS, spes-e-*o*-sus, showy.

**Callistephus,** kal-lis-*tef*-us; from Gr. *kallistos*, most beautiful, and *stephanos*, a crown, referring to the flower. Annuals. The China Aster.

CHINENSIS, tshi-*nen*-sis, of China.

HORTENSIS, hor-*ten*-sis, of gardens.

**Callitriche,** cal-*lit*-rik-e; from Gr. *kalos*, beautiful, and *thrix*, hair—beautiful hair-like (used by Pliny). Submerged Aquatics.

AUTUMNALIS, au-tum-*na*-lis, effective in autumn—this plant grows through autumn and winter.

VERNA, *ver*-na, effective in spring—this plant grows through spring and summer.

**Calluna,** kal-*lu*-na; from Gr. *kalluno*, to cleanse, alluding to the use of this heather as a broom. Shrub.

VULGARIS, vul-*gar*-is, common. Heather or Ling.

**Calocephalus,** kal-o-*sef*-a-lus; from Gr. *kalos*, beautiful, and *kephale*, a head, alluding to the white cord-like stems forming a "beautiful head." Bedding foliage plant.

BROWNII, *brown*-ei, after Brown.

**Calochortus,** kal-o-*kor*-tus; from Gr. *kalos*, beautiful, and *chortus*, grass, referring to the leaves. Half-hardy bulbs. The Mariposa Lily.

ALBA, *al*-ba, white.

CŒRULEUS, se-*ru*-le-us, blue.

LILACINUS, ly-la-*sin*-us, lilac-coloured.

NUTTALLI, nut-*tal*-le, after Nuttall.

PULCHELLUS, pul-*kel*-lus, beautiful.

VENUSTUS, ven-*us*-tus, charming.

**Calopogon,** kal-o-*po*-gon; from Gr. *kalos*, beautiful, and *pogon*, a beard, referring to the fringed floral lip. Terrestrial orchid.

PULCHELLUS, pul-*kel*-lus, beautiful.

**Caltha,** *kal*-tha; from Gr. *kalathos*, a goblet, alluding to the form of the flower. Bog perennials.

LEPTOSEPALA, lep-to-*sep*-a-la, having thin sepals.

PALUSTRIS, pal-*us*-tris, of marshes. The Marsh Marigold.

20

**Caltha** (*continued*)

POLYPETALA, pol-e-*pet*-a-la, many petals. A misleading name, since the flowers have no petals. Their place taken by coloured sepals.

RADICANS, *rad*-e-kans, rooting freely.

**Calycanthus**, kal-i-*kanth*-us; from Gr. *kalyx*, a flower-cup (calyx), and *anthos*, a flower, in reference to the coloured sepals. Shrubs.

FLORIDUS, *flor*-id-us, flowering abundantly.

GLAUCUS, *glaw*-kus, leaves milky-green.

OCCIDENTALIS, ok-sid-en-*ta*-lis, western.

**Calypso**, kal-*ip*-so; called after the ancient Greek goddess of that name Terrestrial orchid.

BOREALIS, bor-e-*a*-lis, northern.

**Calystegia**, kal-is-*te*-je-a; probably from Gr. *kalyx*, a calyx, or cup, and *stege*, a covering, the calyx of some of the Bindweeds being enclosed in two bracts. Twining plants.

HEDERACEA, hed-er-*a*-se-a, ivy-like—the leaf.

PUBESCENS FLORE PLENO, pu-*bes*-senz flor-e pleen-o, downy and double flowered.

SEPIUM, *se*-pe-um, of hedges.

SILVESTRIS, sil-*ves*-tris, of woods.

**Camassia**, kam-*as*-se-a; from Quamash, the N. American Indian name for C. esculenta. Hardy bulbs.

CUSICKII, ku-*sik*-kei, after Cusick.

ESCULENTA, es-ku-*len*-ta, eatable. The Quamash.

LEITCHLINII, lycht-*lin*-ei, after Leitchlin.

**Camellia**, kam-*el*-le-a; after George Joseph Kemel (or Camellus), a Jesuit of Moravia, who travelled in Asia and the East. Greenhouse evergreen flowering shrubs; many named hybrids of C. japonica and other species.

JAPONICA, jap-*on*-ik-a, of Japan.

SANSANQUA, sas-*ang*-kwa, native Japanese name.

THEA, *tee*-a, from *Tsai*, the Chinese name. The Tea Plant.

THEIFERA, tee-*if*-er-a, tea bearing. The Tea Plant.

**Campanula**, kam-*pan*-u-la; from L. *campanula*, a little bell—the Bellflowers. Annuals, biennials, and herbaceous perennials.

ABIETINA, ab-e-*te*-na, of fir woods.

ARVATICA, ar-*vat*-ik-a, from Arvas in Cantabrian Mts.

**Campanula** (*continued*)

BARBATA, bar-*ba*-ta, bearded—the flowers being hairy.

CARPATICA, kar-*pat*-ik-a, Carpathian.

EXCISA, eks-*si*-sa, cut, *i.e.*, the cleft at the base of the bloom segments.

FRAGILIS, *fraj*-il-is, fragile.

GARGANICA, gar-*gan*-ik-a, from Gargano, Italy.

GLOMERATA, glom-er-*a*-ta, clustered—the flowers.

ISOPHYLLA, is-of-*il*-la (or i-so-*fil*-a), equal-leaved.

LACTIFLORA, lak-tif-*lo*-ra, with milk-white flowers.

LATILOBA, lat-e-*lo*-ba, broad-lobed.

MEDIUM, *me*-de-um, middle-sized. The Canterbury Bell.

MURALIS, mu-*ra*-lis, of walls.

PATULA, *pat*-u-la, spreading open—the bells or flowers.

PERSICIFOLIA, per-sis-if-*o*-le-a, peach (Persica)-leaved.

PULLA, *pul*-la, dark-coloured.

PUNCTATA, pung-*ta*-ta, dotted or speckled.

PUSILLA, pu-*sil*-la, dwarf.

PYRAMIDALIS, pir-*am*-id-al-is, pyramidal. The Chimney Bellflower.

RADDEANA, rad-de-*a*-na, Raddean, the Caucasus.

RAPUNCULUS, ra-*pun*-ku-lus, little turnip. The Rampion.

ROTUNDIFOLIA, ro-tun-dif-*o*-le-a, round-leaved. The Harebell.

SARMATICA, sar-*mat*-ik-a, Sarmatian (Poland).

SIBIRICA, si-*bir*-ik-a, Siberian.

TRACHELIUM, trak-*e*-le-um, old generic name for Throatwort.

**Canna**, *kan*-na; from *cana*, Latin name for cane or reed. Tropical herbaceous plants used also for summer bedding. Many hybrids.

INDICA, *in*-dik-a, of India. Indian Shot.

**Cannabis**, *kan*-na-bis; from *kannabis*, Greek name for hemp. An annual plant grown for its seeds and fibres.

SATIVA, *sat*-iv-va, cultivated. The Common Hemp.

**Capsicum**, *kap*-sik-um; from Gr. *kapto*, to bite, referring to the pungency of the fruits. Warm-house annual.

ANNUUM, *an*-nu-um, of annual duration.

**Caragana** kar-a-*gan*-a; from Caragan, the Mongolian name. Shrubs or small trees.

ARBORESCENS, ar-bor-*es*-senz, tree-like.

FRUTESCENS, fru-*tes*-senz, shrubby.

PYGMÆA, *pig*-me-a, dwarf.

SPINOSA, spi-*no*-sa, spiny.

**Carbenia,** kar-*bee*-ne-a; believed to be formed of a compound of first syllables of *Carduus bene*dictus, a synonymous name for the plant on which the present genus was founded.
  BENEDICTA, ben-e-*dik*-ta, blessed. The Blessed Thistle.

**Cardamine,** kar-dam-*i*-ne; from Gr. *cardamon*, watercress, der. rom *kardia*, the heart, and *damao*, to subdue, the plant having properties once used as a heart sedative in medicine. Hardy herbaceous.
  PRATENSIS, pra-*ten*-sis, of meadows. A double form of this cultivated.
  TRIFOLIA, trif-*o*-le-a, three-leaved.

**Carex,** *kar*-eks; the L. name for some kind of Sedge, now applied to the whole group. Grass-like herbs.
  JAPONICA, jap-*on*-ik-a, of Japan.
  PALUDOSA, pal-u-*do*-sa, of marshes.
  PENDULA, *pen*-du-la, drooping.
  RIPARIA re-*pair*-e-a, of river banks.

**Carnegiea,** kar-*nee*-gee-a; after Andrew Carnegie, whose Institute financed collectors of cacti in America. Greenhouse cacti.
  GIGANTEA, ji-*gan*-te-a, of large size. The Giant Cactus of California and Arizona.

**Carpenteria,** kar-pen-*teer*-e-a; named after Professor Carpenter, a botanist, of Louisiana, U.S.A. Shrub.
  CALIFORNICA, kal-if-*or*-nik-a, of California.

**Carpinus,** kar-*pine*-us; ancient L. name for Hornbeam. Deciduous trees.
  BETULUS, *bet*-u-lus, generic name for Birch, which it resembles. The Hornbeam.
  CORDATA, cor-*da*-ta, heart-shaped, the leaves.

**Carpobrotus,** kar-po-*bro*-tus; from Gr. *karpos*, a fruit, and *brotos*, edible, the fruits edible. Greenhouse succulents.
  ACINACIFORMIS, a-sin-*as*-if-*or*-mis, scimitar-shaped, *i.e.*, curved and thick on the outer edge and thin on the inner.
  ɛDULIS, *ed*-u-lis, edible—the fruits. The Hottentot Fig.

**Carum,** *ka*-rum; from Caria, a district in Asia Minor where discovered. Flavouring herbs.
  CARVI (or carui), *kar*-vi, the Caraway.
  PETROSFLINUM, pet-ros-el-*ee*-num, rock parsley. (Gr., *petroselinon*.) The Garden Parsley.

**Caryopteris,** kar-e-*op*-ter-is; from Gr. *karnon*, a nut, and *pteron*, a wing, the fruits being winged. Shrub.
  CLANDONENSIS, clan-don-*en*-sis, of Clandon.
  MASTACANTHUS, mast-a-*kan*-thus, old generic name supposed to be from Gr. *mastax*, a moustache. The Moustache Plant.

**Cassandra,** kas-*san*-dra; a Greek mythological name. Shrub.
  CALYCULATA, kal-ik-u-*la*-ta, small-calyxed.

**Cassia,** *kas*-se-a; from Gr. *kasian*, Greek name of the subject. Greenhouse shrubby plants.
  CORYMBOSA, kor-imb-*o*-sa, corymbose.
  MARYLANDICA, ma-ry-*land*-ik-a, native of State of Maryland.

**Cassinia,** kas-*sin*-e-a; named after Cassini, a French botanist. Shrub.
  FULVIDA, *ful*-vid-a, tawny—the foliage.

**Cassiope,** kas-se-*o*-pe; named after a Queen of Ethiopia, the mother of Andromeda, in Greek mythology. Shrub
  FASTIGIATA, fas-tij-e-*a*-ta, pointed and erct.
  TETRAGONA, tet-rag-*o*-na, four-angled— the leaves in fours.

**Castanea,** kas-*ta*-ne-a; from Gr. *kasta-non*, a chestnut, said to be after Kastana, a district in Thessaly. L. *castanea*, a Chestnut tree. Trees.
  SATIVA, *sat*-iv-a, cultivated, *i.e.*, for crops. The edible Chestnut.

**Catalpa,** kat-*al*-pa; a N. American Indian name for C. bignonioides. Flowering trees.
  BIGNONIOIDES, big-no-ne-*oy*-des, Bignonia-like. The Indian Bean tree.

**Catananche,** kat-an-*ang*-ke; from *katan-anke*, an incentive, alluding to an ancient custom among the Greeks, who used it in love potions. Hardy herbaceous.
  CŒRULEA, se-*ru*-le-a, sky-blue. The Blue Cupidone.

**Cathcartia**, kath-*kar*-te-a; called after J. F. Cathcart, a Judge in the Indian Civil Service. Hardy herbaceous.

VILLOSUS, vil-*lo*-sus, shaggy—with fine hairs.

**Cattleya**, *kat*-le-a; after William Cattley of Barnet, an ardent collector. Tropical orchids.

ACLANDIÆ, ak-*land*-ee-e, after Lady Acland.

BOWRINGIANA, bo-ring-e-*a*-na, after Bowring.

GIGAS, *ji*-gas, giant.

HARRISONÆ, har-ris-*on*-e, Mrs. Harrison's.

INTERMEDIA, in-ter-*med*-e-a, between.

LABIATA, lab-e-*a*-ta, lipped.

MENDELII, men-*del*-ei, after Mendel.

MOSSIÆ, *mos*-ee-e, after Mrs. Moss.

TRIANÆ, tree-*a*-ne, after Dr. Triana.

WARSCEWICZII, war-skew-*ik*-zei, after Warscewicz.

**Ceanothus**, se- (or ke-) an-*o*-thus; ancient Gr. name, supposed to have been applied to a now unknown plant by Theophrastus, the Greek philosopher. Shrubs.

AMERICANUS, am-er-ik-*a*-nus, American.

AZUREUS, az-*ure*-e-us, azure.

DENTATUS, den-*ta*-tus, toothed, the leaf margin.

FLORIBUNDUS, flo-rib-*un*-dus, many-flowered.

INTEGERRIMUS, in-teg-*er*-rim-us, entire, *i.e.*, leaves having unbroken margins.

PAPILLOSUS, pap-il-*o*-sus, pimpled, *i.e.*, the leaves nippled with glands.

RIGIDUS, *rig*-id-us, rigid, stiff.

THYRSIFLORUS, ther-sif-*lo*-rus, flowers in thyrses, *i.e.*, the middle blooms of the cluster having longer stalks than those above and below.

VEITCHIANUS, veech-e-*a*-nus, after Veitch.

**Cedronella**, sed-ron-*el*-la; from Gr. *kedros*; the Cedar of Pliny, the Roman naturalist, doubtless alluding to the fragrance—Balm of Gilead. Aromatic herbs.

TRIPHYLLA, trif-*il*-la, three-leaved.

**Cedrus**, *se*-drus; from ancient Greek name for the Cedar; some suggest name to be derived from Kedron, a river of Judea. Trees.

ATLANTICA, at-*lan*-tik-a, of the Atlas Mountains.

DEODARA, de-o-*dar*-a, from Deodar, an Indian State.

LIBANI, *lib*-an-i, of Mount Lebanon. The Cedar of Lebanon.

**Celastrus**, se-*las*-trus; from Gr. *kelastros*, an evergreen tree. Shrubs, usually climbing.

ARTICULATUS, ar-tik-u-*la*-tus, jointed.

FLAGELLARIS, flaj-el-*lar*-is, whip-like.

SCANDENS, *skan*-dens, climbing.

**Celosia**, se-*lo*-se-a; probably from Gr. *kelos*, burnt, in reference to the burned appearance of the flowers of some species. Greenhouse annuals.

CRISTATA, kris-*ta*-ta, crested. The Cockscomb of florists.

PLUMOSA, plu-*mo*-sa, feathery.

**Celsia**, *sel*-se-a; called after Professor Celsius, of Upsala. Biennials.

ARCTURUS, ark-*tu*-rus, after Arcturus, a yellow star—the colour of the flowers.

CRETICA, *kre*-tik-a, of Crete.

**Centaurea**, sen- (or ken-) *taw*-re-a; from the classical name of a plant (kentaurion or centaureum) in the fables of ancient Greece, which is said to have healed a wound in the foot of Chiron, one of the Centaurs. Annuals and perennials.

BABYLONICA, bab-e-*lon*-ik-a, Babylonian.

CYANUS, sy-*a*-nus, dark blue, Gr. name for the Cornflower. The Cornflower.

DEALBATA, de-al-*ba*-ta, whitened.

GYMNOCARPA, jim-no-*kar*-pa, naked-fruited.

MACROCEPHALA, mak-ro-*sef*-a-la, large-headed.

MONTANA, mon-*ta*-na, of mountains. The Mountain Centaury.

MOSCHATA, mos-*ka*-ta, musky. The Sweet Sultan.

RUTHENICA, ru-*then*-ik-a, Ruthenian.

SUAVEOLENS, swa-ve-ol-enz, sweetly scented. The Sweet Sultan.

**Centradenia**, sen- (or ken-) tra-*de*-ne-a; from Gr. *kentron*, a spur, and *aden*, a gland—spur-like glands on the anthers.

FLORIBUNDA, flor-ib-*un*-da, free-flowering.

**Centranthus**, sen- (or ken-) *tran*-thus; from Gr. *kentron*, a spur, and *anthos*, a flower, the flower having a spur-like base. Hardy herbaceous perennial. Red Valerian.

MACROSIPHON, mak-ro-*sy*-fon, long tubed—the flowers have a long spur.

RUBER, *ru*-ber, red.

**Centropogon**, sen- (or ken-) tro-*po*-gon; from Gr. *kentron*, a spur, and *pogon*, a beard; the fringed or bearded stigma. Greenhouse perennial.

LUCYANUS, loo-se-*a*-nus, after Lucy.

**Cephalaria**, sef- (or kef-) al-*ar*-ea; from Gr. *kephale*, a head—flowers collected into heads. Hardy herbaceous plants.

TATARICA, tah-*tar*-ik-a, of Tartary.

**Cephalotus**, sef- (or kef-) al-*o*-tus; from Gr. *kephalotes*, headed, referring to glandular head of stamens. Greenhouse herbaceous plants.

FOLLICULARIS, fol-lik-ul-*a*-ris, like a follicle—the leaves.

**Cerastium**, ser-*as*-te-um; from Gr. *keras*, a horn, the seed capsules of some appearing like horns as they emerge from the calyx. Rock and border plants.

BIEBERSTEINII, bi-ber-*sti*-nei, after Bieberstein, Russian botanist.

TOMENTOSUM, to-men-*to*-sum, downy. The Snow-in-Summer.

**Cerasus**, ser-*a*-sus, or ser-*as*-us; probably from Cerasunt, a town of Pontus, in Asia, whence the Cherry was first brought to Italy by the Roman hero Lucullus. Now included under Prunus. Flowering and fruiting trees.

AVIUM, *av*-e-um, from L. *avis*, a bird. The Bird Cherry.

JAPONICA, jap-*on*-ik-a, of Japan.

LAUROCERASUS, *law*-ro-ser-*a*-sus, laurel-cherry. The Cherry Laurel or Common Laurel.

LUSITANICA, loo-sit-*a*-nik-a, of Lusitania (Portugal). The Portugal Laurel.

**Ceratostigma**, ser-at-o-*stig*-ma; from Gr. *keras*, a horn, and *stigma*, alluding to the horn-like branches of the stigma. Shrubs.

PLUMBAGINOIDES, plum-ba-gin-*oy*-des, plumbago-like.

WILLMOTTIANA, wil-mot-e-*a*-na, after Miss Willmott.

**Ceratophyllum**, ser-at-o-*fil*-lum; from *keras*, a horn, and *phyllon*, a leaf—the divisions of the leaves suggesting horns. Submerged aquatics. Hornwort.

DEMERSUM, de-*mer*-sum, down under—the water.

SUBMERSUM, sub-*mer*-sum, submerged.

**Cercis**, *ser*-sis; an ancient name given to the Judas tree by the Greek philosopher Theophrastus. Flowering tree.

SILIQUASTRUM, sil-e-*kwas*-trum, from siliqua, a botanic name for pods having a partition between the seeds. The Judas Tree.

**Cereus**, *se*-re-us; from L. but origin uncertain, possibly from L. *cereus*, wax-like or pliant, referring to pliant stems. Greenhouse cacti, with spiny ribs.

FLAGELLIFORMIS, fla-jel-lif-*or*-mis, whip-like.

GRANDIFLORUS, gran-dif-*lo*-rus, large flowered.

MACDONALDIÆ, mak-don-*ald*-ee-e, after Mrs. MacDonald.

NYCTICALUS, nik-*tik*-a-lus, flowering at night.

SPECIOSISSIMUS, spes-e-o-*sis*-se-mus, most handsome.

**Cerinthe**, ser- (or ker-) *in*-thee; from Gr. *keros*, wax, and *anthos*, a flower, referring to flowers frequented by bees. Half-hardy annual. The Honeyworts.

MAJOR, *ma*-jor, larger or greater—the plant.

**Ceropegia**, ker-o-*pe*-je-a; from Gr. *keros*, wax, and *pege*, a fountain, referring to form and waxy appearance of flowers. Greenhouse trailing plants.

ELEGANS, *el*-e-ganz, elegant.

SANDERSONII, san-der-*so*-nei, after Sanderson.

WOODII, *wood*-ei, after Wood.

**Cestrum**, *kes*-trum (or *ses*-trum); Greek name, but applied to another subject. Greenhouse flowering shrubby plants.

AURANTIACUM, aur-*an*-te-ak-um, orange-coloured.

ELEGANS, *el*-e-ganz, elegant.

NEWELLII, new-*el*-ei, after Newell.

**Chamælirion**, kam-e-*lir*-e-on; from Gr. *chamai*, ground, and *leirion*, lily—a dwarf lily-like plant. Herbaceous perennial.

LUTEUM, *lu*-te-um, yellow.

CAROLINIANUM, kar-o-*lin*-e-an-um, of Carolina.

**Chamæpeuce**, kam-e-*pu*-se; from Gr. for small, and *spruce*, from the shape of the leaves. Variegated thistles used for summer bedding.

CASABONÆ, kas-a-*bo*-ne, after Casaubon.

DIACANTHA, di-ak-*an*-tha, having two spines. The Fishbone Thistle.

**Chamærops**, kam-*e*-rops; from Gr. *chamai*, on the ground or dwarf, and *rhops*, a twig, suggesting the dwarfness or twigginess of chamærops in contrast to the tall palms. Greenhouse palms.

EXCELSA, ek-*sel*-sa, lofty.

FORTUNEI, for-*tu*-ne-i, after Fortune.

HUMILIS, *hum*-il-is, dwarf or on the ground.

**Chara,** *kar*-a; origin unknown, possibly from Gr. *karis*, grace (or beauty). Submerged aquatics.

ASPERA, *as*-per-a, rough to the touch.
FRAGIFERA, frag-*if*-er-a, strawberry-like —the red fruiting bodies.
VULGARIS, vul-*gar*-is, common.

**Charieis,** kar-*i*-is; from Gr. *charieis*, elegant, referring to the beauty of the flowers. Hardy annual.

HETEROPHYLLA, het-er-*of*-il-la, varied leaved.

**Cheilanthes,** ky-*lanth*-eez; from Gr. *cheilos*, a lip, and *anthos*, a flower, referring to the indusium. Greenhouse ferns.

ELEGANS, *el*-e-ganz, elegant.
HIRTA ELLISIANA, *hert*-a el-*lis*-e-an-a, Ellis's hairy.
MYRIOPHYLLA, mir-e-*of*-il-a, many of myriad divisions. The Lace Fern.

**Cheiranthus,** ky-*ran*-thus; meaning obscure, probably from Arabic *kheyri* (Cheiri), a name for some sweet-smelling red flower, the same being applied to the Wallflower by the Greeks, their name being derived from *cheir*, a hand, and *anthos*, a flower, the flowers being suitable for hand bouquets. Biennials and perennials.

ALLIONII, al-le-o-nei, of Allioni, Italian botanist.
CHEIRI, *ky*-re, see Cheiranthus above. The Wallflower.

**Cheiridopsis,** ky-rid-*op*-sis; from Gr. *cheiris*, a sleeve, and *opsis*, like, refers to the withered sheath surrounding the new foliage. Greenhouse succulents.

CANDIDISSIMA, kan-did-*is*-sim-a, whitest, very white.
CAROLI-SCHMIDTI, kar-*o*-le-smid-e, after Carl Schmidt.

**Chelidonium,** kel-id-*o*-ne-um; from Gr. *chelidon*, a swallow, the plant (Greater Celandine) being supposed to flower and fade with the arrival and departure of the swallow. Hardy herbaceous.

MAJUS, *ma*-jus, great. There is a double form.

**Chelone,** ke-*lo*-ne; Gr. *kelone*, a tortoise, the helmet of the flower suggesting the shape of that reptile. Herbaceous plants.

BARBATA, bar-*ba*-ta, bearded with hooked hairs.
GLABRA, *glab*-ra, smooth or hairless.
OBLIQUA, ob-*lee*-kwa, with unequal sides.

**Chenopodium,** ken-o-*pod*-e-um; from Gr. *chen*, goose, and *pous*, a foot—the shape of the leaves. The goosefoot family. Hardy culinary plant.

BONUS-HENRICUS, bo-nus-hen-*ree*-kus, good Henry. Good King Henry or Lincolnshire Spinach.

**Chimaphila,** kim-*af*-il-a; from Gr. *cheima*, winter, and *phileo*, to love, the plants are green in winter. Dwarf shrubs.

MACULATA, mak-ul-*a*-ta, spotted. The Spotted Wintergreen.
UMBELLATA, um-bel-*la*-ta, umbelled flowers in umbel-like clusters.

**Chimonanthus,** kim-on-*an*-thus; from Gr. *cheima*, winter, and *anthos*, a flower, alluding to the flowering of the Winter Sweet (C. fragrans) in the early year. Shrub.

FRAGRANS, *fra*-granz, sweet-scented.

**Chionanthus,** ky-on-*an*-thus; from Gr. *chion*, snow, and *anthos*, a flower, the flowers of some species appearing like a fleece of snowflakes. Shrubs.

RETUSA, re-*tu*-sa, blunted—the leaves.
VIRGINICA, vir-*jin*-ik-a, of Virginia.

**Chionodoxa,** ky-on-o-*doks*-a; from Gr. *chion*, snow, and *doxa*, glory—Glory of the Snow. Bulbs.

LUCILLÆ, lu-*sil*-e-e, after Mme. Lucile Boissier.
SARDENSIS, sar-*den*-sis. Sardinian.

**Chionoscilla,** ky-on-o-*sil*-la; a name compounded from *chiono*doxa and *scilla*, the plants being hybrids between these two allied genera. Hardy bulbs.

**Chlidanthus,** kly-*dan*-thus; from Gr. *clideios*, delicate, and *anthos*, a flower. Half-hardy bulbous plants.

FRAGRANS, *fra*-granz, fragrant.

**Chlorophytum,** klo-*rof*-it-um; from Gr. *chloros*, green, and *phyton*, a plant. Greenhouse herbaceous plant with variegated leaves (variety variegatum).

ELATUM, e-*la*-tum, tall growing—the flower stems.

**Choisya,** *choy*-se-a; after M. Choisy, a botanist of Geneva. Shrub.

TERNATA, ter-*na*-ta, with three leaflets. The Mexican Orange.

**Chorizema,** kor-iz-*e*-ma; from Gr. *choros*, a dance, and *zema*, a drink; this name is said to have been given by an early explorer in New Holland (Australia)

**Chorizema** (*continued*)
who had discovered water to his great
joy. Greenhouse flowering shrubs.

CORDATUM SPLENDENS, kor-*da*-tum *splen*-denz, heart-shaped and splendid—
the flowers.

ILICIFOLIUM, il-is-if-*o*-le-um, leaves holly
(Ilex)-like.

**Chrysanthemum**, kris-*an*-the-mum; from
Gr. *chrysos*, gold, and *anthemon*, a
flower, the flowers of many species being
golden, or with yellow disks, as in
Ox-eye Daisy. Annuals and perennials.

ARCTICUM, *ark*-tik-um, Arctic.

CARINATUM, kar-in-*a*-tum, keeled.

CORONARIUM, kor-on-*air*-e-um, crown
or wreath-like.

FRUTESCENS, frut-*es*-senz, shrubby. The
Paris Daisy.

INDICUM, *in*-dik-um, Indian. The green-house Chrysanthemum.

LEUCANTHEMUM, lew-*kan*-the-mum, white
flowered.

MAXIMUM, *maks*-im-um. large. The Large
Ox-eye or Shasta Daisy.

SEGETUM, seg-*e*-tum, of cornfields. The
Corn Marigold.

SINENSE, sin-*en*-se, Chinese. The green-house and border Chrysanthemum.

ULIGINOSUM, u-lij-in-*o*-sum, moisture
loving, *i.e.*, of marshes.

**Chrysogonum**, kris-*og*-o-num; from Gr.
*chrysos*, gold, and *gonu*, a joint, the
golden flowers appearing at the joints
of the stem. Hardy herbaceous.

VIRGINIANUM, vir-*jin*-e-a-num, of Vir-ginia.

**Chrysosplenium**, kris-os-*ple*-ne-um;
from Gr. *chrysos*, gold, and *splen*,
spleen, the colour of the flowers and
medicinal use for the spleen. Dwarf
hardy plants.

ALTERNIFOLIUM, al-tern-if-*o*-le-um, alter-nate leaved.

OPPOSITIFOLIUM, op-pos-it-if-*o*-le-um, op-posite leaved.

**Chysis,** *ky*-sis; from Gr. *chysis*, melt-ing; fused appearance of pollen masses.
Warm-house orchids.

AUREA, *aw*-re-a, golden.

BRACTESCENS, brak-*tes*-enz, with bracts.

**Cichorium**, sik-*or*-e-um, an old Arabic
name for Chicory. Gr. *kichore*. Salad
and root vegetable.

ENDIVIA, en-*dee*-ve-a, endive. The En-dive.

INTYBUS, *in*-tib-us, L. for Endive, the
old name for Chicory. The Chicory.

**Cimicifuga,** sim-is-if-*u*-ga; from L.
*cimex*, a bug, and *fugio*, to run away,
the plant once being used to ward off
fleas. Hence the Eng. name Bugbane.
Hardy herbaceous plants.

DAHURICA, da-*ur*-ik-a, of Dahuri, Asia.

JAPONICA, jap-*on*-ik-a, of Japan.

RACEMOSA, ras-e-*mo*-sa, flowers racemed.

SIMPLEX, *sim*-pleks, simple—the spikes
unbranched.

**Cineraria,** sin-er-*air*-e-a; from L. *cinereus*,
ash-coloured, referring to the grey
down on the undersides of the leaves.
Now included under Senecio.

CRUENTA, kru-*en*-ta, bloody, referring
to the purple colour of the back of
the leaf.

MARITIMA, mar-*it*-im-a, of the sea, or
maritime. The Dusty Miller.

**Cissus,** *sis*-sus; from Gr. *kissos*, ivy, or
from Arabic *qissos*, also indicating ivy.
Greenhouse climbers.

ANTARCTICA, an-*tark*-tik-a, southern.

DISCOLOR, *dis*-kol-or. particoloured, that
is, variegated with silver.

**Cistus,** *sis*-tus; ancient Gr. name of the
plant. The Rock Rose. Flowering shrubs
for the rock garden.

ALBIDUS, *al*-bid-us, nearly white—the
leaves.

CORBARIENSIS, kor-bar-i-*en*-sis, of Cor-bière, S. France.

CRISPUS, *kris*-pus, curly—the leaves
waved.

CYPRIUS, *sip*-re-us, of Cyprus.

FLORENTINUS, flor-en-*te*-nus, of Florence,
Italy.

LADANIFERUS, lad-an-*if*-er-us, yielding
ladanum, a resinous gum. The Gum
Cistus.

LAURIFOLIUS, law-rif-*o*-le-us, laurel-leaved.

MONSPELIENSIS, mon-spe-le-*en*-sis, of
Montpellier.

POPULIFOLIUS, pop-u-lif-*o*-le-us, poplar
(Populus)-leaved.

SALVIFOLIUS, sal-vif-*o*-le-us, sage-leaved.

VILLOSUS, vil-*lo*-sus, hairy.

**Citrus,** *sit*-rus; classical name first ap-plied to another tree. Greenhouse flow-ering and fruiting evergreen shrubs.

AURANTIUM, au-*ran*-te-um, orange. The
Orange.

LIMONIA, li-*mo*-nia, lemon, The Lemon.

MEDICA, *me*-dik-a, median or between.
The Citron.

TRIFOLIATA, trif-ol-e-*a*-ta, three-leaved.

26

**Cladrastis,** klad-*ras*-tis; from ancient Gr. name referring to another tree. Flowering shrubs.

TINCTORIA, tink-*tor*-e-a, of dyers, that is, used by dyers for dyeing.

**Cladium,** *klad*-e-um; from Gr. *klados*, a branch or twig; the flowering stems have twiggy branches. Bog perennial.

MARISCUS, mar-*is*-kus, after the genus Mariscus.

**Clarkia,** *klar*-ke-a; after Capt. Clarke, companion of Lewis, explorer of Rocky Mt. regions. Hardy annuals.

ELEGANS, *el*-e-ganz, elegant.
PULCHELLA, pul-*kel*-la, beautiful, but small.

**Claytonia,** klay-*to*-ne-a; after John Clayton, an American plant collector. Mostly annuals.

PERFOLIATA, per-fol-e-*a*-ta, the stem appearing to pass through the leaf.
SIBIRICA, si-*bir*-ik-a, of Siberia.

**Clematis,** *klem*-a-tis; from Gr. *klema*, a vine branch, alluding to the vine-like habit of the climbing species. Herbaceous perennials and climbing shrubs.

COCCINEA, kok-*sin*-e-a, scarlet.
DAVIDIANA, da-vid-i-*a*-na, after David, plant collector.
FLAMMULA, *flam*-mul-a, a little flame.
FLORIDA, *flor*-id-a, flowering richly.
GRAVEOLENS, *grav*-e-o-lenz, strong-smelling.
LANUGINOSA, lan-u-jin-*o*-sa, downy.
MONTANA, mon-*ta*-na, of mountains.
ORIENTALIS, or-i-en-*ta*-lis, eastern.
PANICULATA, pan-ik-ul-*a*-ta, panicled.
RECTA, *rek*-ta, upright.
VITALBA, vit-*al*-ba, the white vine. The Wild Clematis, Virgin's Bower or Traveller's Joy.
VITICELLA, vit-e-*sel*-la, vine-bower.

**Cleome,** kle-o-me; derivation uncertain. Tropical plants.

GIGANTEA, ji-*gan*-te-a, large.

**Clerodendron,** kler-o-*den*-dron; said to be from Gr. *kleros*, chance, and *dendron*, a tree, of uncertain significance. Greenhouse and hardy shrubs.

BALFOURI, bal-*four*-i, after Balfour.
FALLAX, *fal*-laks, false or deceitful.
FARGESII, *far*-ge-sei, after Farges, a shrub collector.
FŒTIDUM, *feet*-id-um, fetid or strong smelling.
FRAGRANS, *fra*-granz, fragrant.
TRICOTOMUM, trik-*ot*-om-um, with three points—the leaves.

**Clethra,** *kle*-thra; from *klethra*, the Greek name for an Alder, which some species resemble. Shrubs and trees.

ACUMINATA, a-ku-min-*a*-ta, long-pointed —the leaves.
ALNIFOLIA, al-nif-*o*-le-a, alder (Alnus)-leaved.

**Clianthus,** kli-*an*-thus; from Gr. *kleios*, glory, and *anthos*, a flower. Climbing shrubs.

PUNICEUS, pu-*nis*-e-us, red or purple. New Zealand Glory Pea.

**Clintonia,** klin-*to*-ne-a; after De Witt Clinton, Governor of New York, at the time of Douglas, the plant collector. Woodland herbs.

UNIFLORA, u-nif-*lor*-a, one-flowered.

**Clivia,** *kly*-ve-a; named after a Duchess of Northumberland, member of the Clive family.

MINIATA, min-e-*a*-ta, vermilion-coloured. Many hybrids from this in yellow to orange scarlet.

**Cnicus,** *ny*-kus; L. name of safflower, early applied to thistles. Variegated thistles used for summer bedding.

CASABONÆ, kas-ab-*o*-ne, after Casabona.
DIACANTHA, di-ak-*an*-tha, two spined. The Fishbone Thistle.

**Cobæa,** ko-*be*-a; after Barnadez Cobo, a Spanish naturalist in Mexico, whence most species come. Half-hardy climbing plants.

SCANDENS, *skan*-denz, climbing. The Cup-and-Saucer Flower.

**Cochlearia,** kok-le-*ar*-ea; from L. *cochlea*, a spoon; refers to concave leaves of Scurvy Grass (C. officinalis).

ARMORACIA, ar-mo-*ra*-se-a, one-time generic name.

**Cocos,** ko-kos; from Portuguese *coco*, a monkey, in allusion to end of nut simulating a monkey's head. Greenhouse palms.

NUCIFERA, nu-*sif*-er-a, nut-bearing.
WEDDELLIANA, wed-del-li-*a*-na, after Weddell.

**Codiæum,** ko-de-*ee*-um; probably from Gr. for head, the leaves being used for wreaths. Tropical variegated-leaved shrubs; numerous garden hybrids.

ANGUSTIFOLIUM, an-gus-tif-*o*-le-um, narrow leaved.
CHELSONI, *chel*-son-e, after Chelson.

27

**Codiæum** (*continued*)

INTERRUPTUM, in-ter-*rup*-tum, interrupted, the leaf being reduced to midrib only in places along its length.

PICTURATA, pik-tur-*a*-ta, coloured, picture-like.

VARIEGATUM, var-e-eg-*a*-tum, variegated leaved.

WARRENII, war-*ren*-e, after Warren.

**Codonopsis**, ko-don-*op*-sis; from Gr. *kodon*, a bell, and *opsis*, resemblance, the flowers being bell-shaped. Tender perennials.

CLEMATIDEA, klem-at-*id*-e-a, clematis-like.

OVATA, o-*va*-ta, egg-shaped—the leaves.

**Cœlogyne**, se-*log*-e-ne; from Gr. *koilos*, hollow, and *gyne*, a woman, in allusion to the pistil. Cool-house orchids.

CRISTATA, kris-*ta*-ta, crested lipped.

DAYANA, day-*a*-na, after Day.

MASSANGEANA, mas-san-ge-*a*-na, after Massange.

SPECIOSA, spes-e-*o*-sa, showy.

**Coffea**, *kof*-fe-a; from *quahouch*, the Arabic name for the liquor. Tropical shrub.

ARABICA, ar-*ab*-ik-a, of Arabia. The Coffee Tree.

**Coix**, *ko*-iks; Theophrastus's name for a reed-leaved plant. Annual ornamental grass.

LACHRYMA, *lak*-rim-a, a tear, the large grey seeds being likened to tear drops. Job's Tears.

**Colchicum**, *kol*-chi-kum; named after Colchis, a province of Asia Minor, where the plant abounds. Hardy bulbs.

AUTUMNALE, au-tum-*na*-le, flowering in autumn.

BYZANTINUM, bi-zan-*te*-num, from Byzantium.

GIGANTEUM, ji-*gan*-te-um, gigantic.

SPECIOSUM, spes-e-*o*-sum, showy.

**Coleus**, *kol*-e-us; from Gr. *koleos*, a sheath, in allusion to the combined stamens. Greenhouse foliage plants; many garden hybrids.

BLUMEI, *blu*-me-i, after Blume.

FREDERICI, fred-er-*e*-ki, after Frederic.

THYRSOIDEUS, thyr-so-*id*-e-us, thyrse-like —the flower spikes.

VERSCHAFFELTII, ver-shaf-*felt*-ei, after Verschaffelt.

**Colletia**, kol-*le*-te-a; called after M. Collet, a French botanist. Shrubs.

ARMATA, ar-*ma*-ta, armed—with thorns.

CRUCIATA, kru-se-*a*-ta, crossed—the spines.

**Collinsia**, kol-*lin*-se-a; called after Zaccheus Collins, a naturalist, of Philadelphia. Hardy annuals.

BICOLOR, *bik*-o-lor, two-coloured.

**Collomia**, kol-*lo*-me-a; from *kolla*, glue, referring to mucilage around the seeds. Hardy annual.

COCCINEA, kok-*sin*-e-a, scarlet.

**Colocasia**, kol-o-*ka*-se-a; from Gr. *kolokasia*, Greek name for the root of an Egyptian plant. Tropical tuberous-rooted foliage plants.

ANTIQUORUM, an-te-*kor*-um, ancient.

ESCULENTUM, es-kul-*en*-tum, esculent or edible.

**Columnea**, kol-*um*-ne-a; after Fabius Columna, Italian writer on plants. Tropical trailers.

GLORIOSA, glor-e-*o*-sa, glorious.

MAGNIFICA, mag-*nif*-ik-a, magnificent.

**Colutea**, ko-*lu*-tea; derivation obscure, appears to have been adopted from *koloutea*, a name used by Theophrastus. Shrubs.

ARBORESCENS, ar-bor-*es*-sens, tree-like.

CRUENTA, kru-*en*-ta, blood-red.

MEDIA, *me*-de-a, medium.

**Commelina**, kom-me-*le*-na; named after J. and K. Commelin, early Dutch botanists. Annuals and perennials.

CŒLESTIS, se-*les*-tis, sky-blue.

TUBEROSA, tu-ber-*o*-sa, tuberous rooted.

**Conicosia**, kon-e-*ko*-se-a; from Gr. *konikos*, conical, the conical top of the fruit. Greenhouse succulent.

PUGIONIFORMIS, pu-je-o-nif-*or*-mis, dagger-shaped.

**Conophytum**, kon-*of*-i-tum; from Gr. *konos*, a cone, and *phyton*, a plant—in allusion to the form of many species— an inverted cone. Greenhouse succulents. The Pebble Plants.

AGGREGATUM, ag-greg-*a*-tum, clustered.

GLOBOSUM, glob-*o*-sum, spherical.

LONGUM, *long*-um, long—the leaves, for the genus.

MINUTUM, min-*u*-tum, small.

MUNDUM, *mund*-um, from L. *mundus*, of the earth—plant grows close to the soil.

**Convallaria**, kon-val-*lair*-e-a; from L. *convallium*, a valley, the natural habitat of the Lily of the Valley. Hardy herbaceous.

MAJALIS, maj-*a*-lis, May—time of flowering.

28

**Convolvulus,** kon-*vol*-vu-lus; from L. *convolvo*, to entwine, alluding to the twining habit of some species. Shrubby, herbaceous and rock plants.

ALTHÆOIDES, al-the-*oy*-des, like an Althæa (hollyhock)—the flowers.

CANTABRICA, kan-*tab*-rik-a, Cantabria, Spain.

CNEORUM, ne-*or*-um, meaning obscure; from Gr. *kneoron*. The genus Cneorum.

LINEATUS, lin-e-*a*-tus, with lines.

MAURITANICUS, maw-re-*tan*-ik-us, of Mauretania (Morocco).

NITIDUS, *nit*-id-us, somewhat glossy.

SOLDANELLA, sol-dan-*el*-la, leaves like a Soldanella.

TENUISSIMUS, ten-u-*is*-sim-us, most slender.

TRICOLOR, *trik*-o-lor, three-coloured.

**Coprosma,** kop-*roz*-ma; from Gr. *kopros*, dung, and *osme*, a smell; alluding to unpleasant odour. Greenhouse shrub.

BAURI VARIEGATA, bo-*ee*-re var-e-eg-*a*-ta, Bauer's variegated-leaved.

**Cordyline,** kor-dil-*y*-ne; from Gr. *kordyle*, a club. Greenhouse palm-like plants.

AUSTRALIS, aus-*tra*-lis, southern.

INDIVISA, in-de-*vy*-sa, not divided.

LINEATA, lin-e-*a*-ta, lined.

**Coreopsis,** kor-e-*op*-sis, from Gr. *koris*, a bug or tick, and *opsis*, a resemblance, from the appearance of the seed, hence Eng. name of Tickseed. Annuals and perennials.

ARISTOSA, ar-is-*to*-sa, bearded—the seed.

ATKINSONII, at-kin-*so*-nei, after Atkinson.

AURICULATA, aur-ik-ul-*a*-ta, ear-shaped —the leaves.

CORONATA, kor-on-*a*-ta, crowned.

DRUMMONDII, drum-*mon*-dei, after Drummond.

GRANDIFLORA, gran-dif-*lo*-ra, large-flowered.

LANCEOLATA, lan-se-o-*la*-ta, small lance-shaped—the leaves.

ROSEA, *ro*-ze-a, rosy.

TINCTORIA, tink-*tor*-e-a, ref. to dyeing—the variously coloured flowers.

VERTICILLATA, ver-tis-il-*la*-ta, whorled——the leaves.

**Coriaria,** kor-e-*a!r*-e-a; from L. *corium*, hide, probably from the use made of some kinds in tanning leather. The Tanner's tree. Shrubs.

TERMINALIS, ter-min-*a*-lis, terminal—the flowers.

**Cornus,** *kor*-nus; from L. *cornus*, a horn, from the hard nature of the wood. Trees and shrubs.

ALBA, *al*-ba, white.

CANADENSIS, kan-a-*den*-sis, of Canada.

CANDIDISSIMA, kan-did-*is*-sim-a, very white—the flowers.

CAPITATA, kap-it-*a*-ta, headed—grouping of flowers.

FLORIDA, *flor*-id-a, flowering richly.

FRAGIFERA, frag-*if*-er-a, strawberry-like —the fruits.

GLABRATA, gla-*bra*-ta, glabrous.

KOUSA, *koo*-sa, a Japanese name.

MAS, mas, male (*mascula* of Linnæus).

NUTTALLII, *nut*-al-lei, after Nuttall.

SANGUINEA, san-*gwin*-e-a, blood-red—the twigs.

**Corokia,** kor-*o*-ke-a; adapted from Korokia, the Maori name. Shrubs.

BUDDLEOIDES, bud-le-*oy*-des, buddleia-like.

COTONEASTER, ko-to-ne-*as*-ter, Cotoneaster-like.

**Coronilla,** kor-o-*nil*-la; meaning a little crown, from L. *corona*, a crown or garland, in reference to the disposition of the flowers. Rock plants and shrubs.

CAPPADOCICA, kap-pa-*do*-sik-a, Cappadocian, Asia Minor.

EMEROIDES, em-er-*oy*-des, emerus-like.

EMERUS, *em*-er-us, old generic name, meaning cultivated.

GLAUCA, *glaw*-ka, glaucous.

MINIMA, *min*-e-ma, smallest.

VARIA, *var*-e-a, varying in colour.

**Correa,** kor-*re*-a; after Jose Correa de Serra, a Portuguese botanist. Greenhouse evergreen flowering shrubs.

CARDINALIS, kar-din-*a*-lis, red.

SPECIOSA, spes-e-*o*-sa, showy.

VENTRICOSA, ven-trik-*o*-sa, inflated.

**Cortusa,** kor-tu-za; after J. A. Cortusus, Italian botanist. Hardy alpine perennial.

MATTHIOLII, mat-*the*-ol-ei, after Matthiol.

**Corydalis,** kor-*id*-a-lis; from Gr. *korydalis*, a crested lark (der., *korys*, a helmet), alluding to the shape of the petals. Herbaceous and rock plants.

BULBOSA, bul-*bo*-sa, bulbous.

CAVA, *ka*-va, hollow, or cave-like—the bulbous root.

CHEILANTHIFOLIA, ky-lanth-e-*fo*-le-a, leaved like Cheilanthus (Lace fern).

LUTEA, *lu*-te-a, yellow.

SOLIDA, *sol*-id-a, solid—the fleshy root.

THALICTRIFOLIA, thal-ik-trif·*o*-le-a, Thalictrum-leaved.

**Corylopsis**, kor-il-*op*-sis; from Gr. *korylos*, hazel, and *opsis*, like—hazel-like (as to foliage).

    PAUCIFLORA, paw-se-*flor*-a, few-flowered.
    SPICATA, spe-*ka*-ta, flowers in spikes.

**Coryphantha**, kor-if-*anth*-a; from Gr. *koryphe*, top, and *anthos*, a flower, the flowers appearing at the top of the plant. Cacti.

    CHLORANTHA, klor-anth-a, greenish yellow.
    CLAVA, *kla*-va, club-shaped.
    ELEPHANTIDENS, el-ef-*an*-tid-enz, elephant's tooth—the big mamæ.
    MACROMERIS, mak-rom-e-ris, large flowered.
    OTTONIS, ot-*to*-nis, after Ottoni.
    PECTINATA, pek-tin-*a*-ta, comb-like.
    PYCNACANTHA, pik-nak-*an*-tha, densely spined.
    SULCOLANATA, sul-col-an-*a*-ta, woolly groved.

**Corylus**, *kor*-il-us; possibly from Gr. *korys*, a hood or helmet, *i.e.*, the calyx covering the nut; or from Gr. *karyon*, a hazel nut. Shrubs.

    AVELLANA, av-el-*la*-na, after Avella, a town of Campania, where the Hazel was largely grown for its nuts. The Hazel, Cobnut and Filbert.

**Cosmea**, *kos*-me-a; from Gr. *kosmos*, beautiful. A synonym of Cosmos, which see.

**Cosmos**, *kos*-mos; from Gr. *kosmos*, beautiful, in reference to the flowers. Half-hardy annual

    BIPINNATUS, bip-in-*a*-tus, the leaves doubly pinnate.
    SULPHUREUS, sul-*fur*-e-us, sulphur coloured—the flowers.

**Cotoneaster**, ko-to-ne-*as*-ter; from *cotoneus*, old Latin name for a quince, and *aster*, probably a corruption of *ad instar*, meaning a likeness, *i.e.*, quince-like. Shrubs.

    ADPRESSA, ad-*pres*-sa, close, pressed-down growth; or fruits closely pressed against the branch.
    APPLANATA, ap-lan-*a*-ta, the branches plane-line, or flat.
    BULLATA, bul-*la*-ta, wrinkled—the leaves.
    BUXIFOLIA, buks-e-*fo*-le-a, box-(Buxus)-leaved.
    CONGESTA, kon-*jes*-ta, crowded—the habit.
    DIVARICATA, di-var-e-*ka*-ta, spread-out, forking—the branches.
    FRANCHETII, fran-*schet*-ei, after Franchet.

**Cotoneaster** (*continued*)

    FRIGIDA, *frij*-id-a, cold, frosty, probably native habitat.
    HARROVIANA, har-ro-ve-*a*-na, after G. Harrow, once of Coombe Wood.
    HENRYANA, hen-re-*a*-na, after Dr. Augustine Henry.
    HORIZONTALIS, hor-e-zon-*ta*-lis, horizontal—habit of growth.
    HUMIFUSA, hum-e-*fu*-sa, spread on the ground.
    LUCIDA, *lu*-sid-a, shining—the leaves.
    MICROPHYLLA, mi-krof-*il*-la, small-leaved.
    MULTIFLORA, mul-tif-*lo*-ra, many-flowered.
    PANNOSA, pan-*no*-sa, woolly—the foliage.
    ROTUNDIFOLIA, ro-tun-dif-o-le-a, round-leaved.
    SALICIFOLIA, sal-is-e-*fo*-le-a, willow (Salix)-leaved.
    SIMONSII, *si*-mons-ei, after Simons.

**Cotula**, *kot*-u-la; from Gr. *kotyle*, a cup, possibly in allusion to the cup-like heads of some of the species. Dwarf creeping plants.

    BARBATA, bar-*ba*-ta, having hooked hairs.
    SQUALIDA, *skwol*-id-a, squalid or lowly—flower-heads dingy.

**Cotyledon**, kot-il-e-don; from Gr. *kotyle*, a cup, from the round, concave leaves of the Navelwort, C. umbilicus. Mainly greenhouse succulent perennials.

    AGAVOIDES, ag-a-*voy*-dez, agave-like.
    ATROPURPUREA, *a*-tro-pur-*pur*-e-a, dark purple.
    FULGENS, *ful*-jenz, shining or glowing.
    GIBBIFLORA, gib-bif-*lo*-ra, humped flowers.
    METALLICA, me-*tal*-lik-a, metallic—colouring of foliage.
    RETUSA, ret-*u*-sa, blunt leaved.
    SECUNDA, sek-*un*-da, one-sided—the flower spike.
    SIMPLICIFOLIA, sim-plis-e-*fo*-le-a, simple, not compound, leaves.
    UMBILICUS, um-bil-e-kus or um-*bil*-e-kus, navel-like centre of leaves.

**Crambe**, *kram*-bee; from Gr. *krambe*, cabbage. Hardy herbaceous plants.

    CORDIFOLIA, kor-dif-*o*-le-a, heart-shaped leaves.
    ORIENTALIS, or-e-en-*ta*-lis, Oriental or Eastern.
    MARITIMA, mar-*it*-im-a, sea coast or maritime. The Sea Kale.

**Crassula**, *cras*-sul-a; from L. *crassus*, thick, referring to the thick or fleshy leaves. Greenhouse succulents.

    COCCINEA, kok-*sin*-e-a, scarlet.

**Crassula** (*continued*)

COTYLEDON, kot-il-*e*-don, cotyledon-like.

FALCATA, fal-*ka*-ta, scythe-shaped—the leaves.

LACTEA, *lak*-te-a, milk white.

**Cratægus**, kra-*te*-gus; Gr. *krataigos*, a flowering thorn, believed to be derived from *kratos*, strength, alluding to the hardness of the wood. Shrubs and trees.

COCCINEA, kok-*sin*-e-a, scarlet—the fruits.

CORDATA, kor-*da*-ta, heart-shaped—the leaves.

CRUS-GALLI, kroos-*gal*-le, lit. a cock's leg. The Cockspur Thorn.

MACRANTHA, mak-*ran*-tha, large spines.

MOLLIS, *mol*-lis, downy.

MONOGYNA, mon-o-*jin*-a, having a single pistil.

ORIENTALIS, or-e-en-*ta*-lis, eastern

OXYACANTHA, oks-e-a-*kan*-tha, sharp thorn. The Hawthorn or May.

PUNCTATA, punk-*ta*-ta, speckled—the fruits.

TOMENTOSA, to-men-*to*-sa, downy.

**Crepis**, *kre*-pis; from Gr. *krepis*, a sandal, reference not clear. Hardy annuals.

BARBATA, bar-*ba*-ta, bearded.

RUBRA, *roo*-bra; red or ruddy.

**Crinum**, *kry*-num; from Gr. *krinon*, a lily. Greenhouse and hardy bulbous plants.

CAPENSE, ka-*pen*-se, of the Cape of Good Hope.

MOOREI, *moor*-e-e, after Moore.

POWELLII, *pow*-el-ei, after Powell.

**Crocus**, *kro*-kus; Gr. *krokos*, saffron, probably from *kroke*, a thread, the filaments of the styles being the source of the dye. Bulbous perennials.

AUREUS, *aw*-re-us, golden.

BIFLORUS, bif-*lor*-us, two-flowered. The Scotch Crocus.

CHRYSANTHUS, kris-*anth*-us, golden-flowered.

MINIMUS, *min*-e-mus, smallest.

NUDIFLORUS, nu-dif-*lor*-us, naked-flowered—no foliage present.

OCHROLEUCUS, ok-ro-*lew*-kus, yellowish-white.

SATIVUS, *sat*-iv-us, cultivated. The Saffron Crocus.

SIEBERI, *si*-ber-e, after Sieber, a botanist.

SUSIANUS, soos-e-*a*-nus, from Susa, Persia.

VERNUS, *ver*-nus, spring flowering. The Dutch or spring crocuses are derived from this species.

VERSICOLOR, ver-*sik*-o-lor, changing or varied colour.

**Crossandra**, kros-*san*-dra; from Gr. *krossus*, a fringe and *aner*, a man—fringed anthers. Tropical flowering shrubs.

UNDULÆFOLIA, un-dul-e-*fol*-e-a, wavy foliage.

**Crucianella**, kru-se-an-*el*-la; from L. *crux*, a cross, lit. a little cross—the Crossworts. Herbaceous and rock plants.

STYLOSA, sty-*los*-a, large styled.

**Cryptogramme**, krip-to-*gram*-me; from Gr. *kryptos*, hidden, and *gramme*, a line, the place of fruiting (spores) being concealed. Hardy fern.

CRISPA, *kris*-pa, curled. The Parsley Fern.

**Cryptocoryne**, krip-to-*kor*-in-e; from Gr. *kryptos*, hidden, and *koryne*, a club—the club-like spadix is enclosed in the spathe. Tropical submerged aquatics.

CILIATA, sil-e-*a*-ta, fringed with fine hairs—the spathe.

CORDATA, kor-*da*-ta, heart-shaped—the leaves.

**Cryophytum**, kry-*of*-e-tum; from Gr. *kyros*, ice, and *phyton*, a plant, alluding to the ice-like appearance of the plant.

CRYSTALLINUM, kris-tal-*le*-num, crystal-like. The Ice Plant.

**Cryptomeria**, krip-to-*meer*-e-a; from Gr. *kryptos*, hidden, and *meros*, a part, alluding to latent characters, or parts of the tree not being easily understood. Coniferous trees.

ELEGANS, *el*-e-ganz, elegant.

JAPONICA, jap-*on*-ik-a, of Japan.

**Cucumis**, *ku*-ku-mis, from L. *cucumis*, the cucumber. Greenhouse climbing plants.

CITRULLUS, sit-*rul*-lus, Citrus or Citron-like. The Water Melon.

MELO, *mee*-lo, melon. The Melon.

SATIVUS, *sat*-iv-us, cultivated. The Cucumber.

**Cucurbita**, ku-*ker*-bit-a; from L. *curbita*, a gourd. Greenhouse and tender climbing plants. Gourds and pumpkins.

OVIFERA, o-*vif*-er-a, egg-like.

PEPO, *peep*-o, the pre-Linnæan name. The Vegetable Marrow.

**Cuphea**, *ku*-fe-a; from Gr. *kuphos*, curved, the form of the seed pods. Greenhouse and bedding plants.

IGNEA, *ig*-nea, fiery—colour of flowers. The Cigar Plant.

MINIATA, min-e-*a*-ta, vermilion coloured.

PLATYCENTRA, plat-y-*ken*-tra, broad.

**Cupressus,** ku-*pres*-sus; classical name, said to be from Gr. *kuo*, to produce, *parisos*, equal, alluding to the symmetrical form of Italian Cypress; possibly from an ancient Latin word signifying a box, the wood once being used for coffins. Coniferous trees—the Cypresses.

ARIZONICA, ar-i-*zon*-ik-a, of Arizona, U.S.A.

FUNEBRIS, fu-*ne*-bris, pertaining to funerals.

LAWSONIANA, law-so-ne-*a*-na, after Lawson; an Edinburgh nurseryman.

LUSITANICA, lu-sit-*a*-nik-a, of Lusitania (Portugal).

MACROCARPA, mak-ro-*kar*-pa, large-fruited.

NOOTKATENSIS, noot-ka-*ten*-sis, of Nootka, N. America.

OBTUSA, ob-*tu*-sa, blunt—the leaves.

PISIFERA, pis-*if*-er-a, pea-bearing, the cones being round like peas.

SEMPERVIRENS, sem-per-*veer*-enz, always green, Italian Cypress.

**Cyananthus,** sy-an-*an*-thus; from Gr. *kyanos*, dark blue, and *anthos*, a flower. Rock plants.

INCANUS, in-*ka*-nus, hoary, with white hairs.

**Cyathea,** sy-a-*the*-a; from Gr. *kyatheion*, a little cup, the shape of the spore cases. Greenhouse tree fern.

DEALBATA, de-al-*ba*-ta, white-washed—the colour of the under surface of the frond.

**Cycas,** *sy*-kas; from Gr. *kykas*, the name of a palm tree. Tropical foliage (palm-like) plants.

REVOLUTA, rev-ol-*u*-ta, rolled back.

**Cyclamen,** *sik*-la-men; name a contraction of Gr. *kyklaminos*, from *kyklos*, a circle, alluding to the coiled stem of the seed vessel. Herbaceous plants.

CILICICUM, sil-*is*-ik-um, Cicilian.

COUM, *coo*-um, of Cous or Cos, an island off Turkey.

EUROPÆUM, u-ro-*pe*-um, European.

HEDERÆFOLIUM, hed-er-e-*fo*-le-um, ivy (Hedera)-leaved.

IBERICUM, i-*ber*-ik-um, of Iberia, Transcaucasia.

LATIFOLIUM, lat-if-*ol*-e-um, broad-leaved.

NEAPOLITANUM, ne-a-pol-e-*ta*-num, of Naples.

PERSICUM, *per*-sik-um, of Persia.

REPANDUM, re-*pan*-dum, scalloped, the leaf margins.

**Cydonia,** sy-*do*-ne-a; name given by the ancients to the common quince, which grew in abundance at Cydon, in Crete. Trees and shrubs.

JAPONICA, jap-*on*-ik-a, Japanese.

MAULEI, *mawl*-ei, after Maule, of Bristol.

VULGARIS, vul-*gar*-is, common. The Quince.

**Cymbidium,** sim-*bid*-e-um; from Gr. *kymbe*, a boat, referring to a hollow in the lip. Cool-house orchids.

EBURNEUM, eb-*ur*-ne-um, like ivory.

LOWIANUM, low-e-*an*-um, after Low.

TRACEYANUM, tra-cy·*a*-num, after Tracey.

**Cynara,** sin-*ar*-a; from Gr. *kyon*, a dog, with reference to spines on involucre suggesting dog's teeth.

CARDUNCULUS, kar-*dun*-kul-us, Spanish artichoke. The Cardoon.

SCOLYMUS, *skol*-im-us, after or like Scolymus. The Globe Artichoke.

**Cynoglossum,** sin-o-*glos*-sum; from Gr. *kyon*, a dog, and *glossa*, a tongue. The Hound's-tongue. Annuals, biennials and perennials.

AMABILE, am-*a*-bil-e, lovely.

CŒLESTINUM, se-les-*te*-num, heavenly blue.

WALLICHII, *wol*-lich-ei, after Wallich.

**Cyperus,** *sy*-per-us; from old Greek name (*kypeiros*) for a sedge. Waterside perennials.

ALTERNIFOLIUS, al-ter-nif-*o*-le-us, alternate leaved.

LONGUS, *long*-us, long, tall.

PAPYRUS, pa-*py*-rus, paper. The Egyptian Papyrus.

**Cyphomandra,** ky-fom-*an*-dra; or sif-o-*man*-dra; from Gr. *kyphoma*, a hump, and *aner*, a man—the anthers form a hump. Greenhouse fruiting shrub.

BETACEA, be-*ta*-ce-a, beet-like or esculent. The Tree Tomato.

**Cypripedium,** sip-re-*pe*-de-um; from Gr. *Kypris*, one of the names of Venus, and *podion*, a little foot or slipper, literally Venus's Slipper, now changed into Lady's or Ladies' Slipper. Greenhouse and hardy terrestrial orchids.

BELLATULUM, bel-*lat*-u-lum, pretty.

CALCEOLUS, kal-se-*o*-lus, a little slipper.

INSIGNE, in-*sig*-ne, striking.

MACRANTHUM, mak-*ranth*-um, long-flowered.

REGINÆ, re-*ji*-ne, of the queen—Queen Victoria.

**Cypripedium** (*continued*)

SPECTABILE, spek-*tab*-il-e, remarkable, showy. The Canadian Moccasin Flower.

SPICERIANUM, spi-ser-e-*a*-num, after Spicer.

SUPERBIENS, sup-*erb*-e-enz, superb.

VENUSTUM, ven-*us*-tum, charming.

VILLOSUM, vil-*lo*-sum, shaggy.

**Cyrtanthus**, kyr- (or ser-) *tan*-thus; from Gr. *kyntos*, curved, and *anthos*, a flower; the flowers bend down from the main stalk. Greenhouse bulbs.

ANGUSTIFOLIUS, an-gus-tif-*ol*-e-us, narrow leaved.

LUTESCENS, lu-*tes*-cenz, becoming yellow.

MACKENII, mak-*ken*-ei, after Macken.

MACOWANII, mak-*kow*-an-ei, after Macowan.

**Cystopteris**, sist- (or kist-) *op*-ter-is; from Gr. *kystis*, a bag, and *pteris*, a fern, alluding to the sac-like covering of the sori, Bladder Fern. Hardy ferns.

BULBIFERUM, bul-*bif*-er-um, ref. to little green balls or buds under fronds.

FRAGILIS, *fraj*-il-is, fragile—finely leaved.

MONTANA, mon-*ta*-na, of mountains.

**Cytisus**, *sy*-tis-us; Gr. *kytisos*, trefoil, pertaining to the leaves of many species; said to be derived from Cythnos, an island of the Ægean, where some species abound. Also pronounced *sit*-e-sus. Shrubs.

ALBUS, *al*-bus, white.

ARDOINII, ar-*do*-in-ei, after Ardoin, who discovered it.

DALLIMOREI, dal-le-*more*-ei, after Dallimore.

DECUMBENS, de-*kum*-bens, prostrate.

FRAGRANS, *fra*-granz, fragrant.

HIRSUTUS, hir-*su*-tus, hairy.

KEWENSIS, kew-*en*-sis, of Kew Gardens.

LEUCANTHUS, lu-*kan*-thus, white-flowered.

NIGRICANS, *nig*-re-kans, turning black—the faded flowers.

PURGANS, *pur*-gans, purging—its use in medicine.

PURPUREUS, pur-*pur*-e-us, purple.

SCOPARIUS, sko-*par*-e-us, from L. *scopæ*, a broom, a broom made of twigs. The Common Broom.

SESSILIFOLIUS, ses-sil-if-*o*-le-us, sessile-leaved, *i.e.*, stalkless.

**Daboecia**, da-bo-*e*-se-a; from its Irish name, St. Dabeoc's Heath. Dwarf shrub.

POLIFOLIA, pol-if-*o*-le-a, L. *polio*, to polish, and *folium*, a leaf, smooth-

**Daboecia** (*continued*)

leaved. Some authorities contend the word means *polium*, Germander-leaved.

**Dactylis**, *dak*-til-is; from Gr. *daktylos*, a finger, from the shape of the panicle. The Cocksfoot Grass. Ornamental grass.

GLOMERATA, glom-er-*a*-ta, clustered. Variegated form, elegantissima, grown in gardens.

**Dahlia**, *dah*-le-a; named after Andreas Dahl, a Swedish botanist and pupil of Linnæus. Originally (and still in America) pronounced *dah*-le-a and not *day*-le-a as commonly heard. All dahlias usually grown are hybrids from species rarely seen in gardens. Half-hardy tuberous herbaceous plants.

COCCINEA, kok-*sin*-e-a, scarlet.

IMPERIALIS, im-pcer-e-*a*-lis, powerful.

JUAREZII, ju-a-*re*-zei, after Juarez.

MERCKII, *merk*-ei, after Merck.

VARIABILIS, var-e-*ab*-il-is, variable coloured.

ZIMAPANII, zim-ap-*a*-nei, after Zimapan.

**Dalea**, *day*-le-a; after Dr. Samuel Dale, English botanist. Hardy annual and greenhouse sub-shrub.

ALOPECUROIDES, al-o-pek-u-*roy*-dez, like Alopecurus (Foxtail grass).

MUTABILIS, mu-*ta*-bil-is, changeable.

**Danæ**, *dan*-ee; name of a daughter of King Acrisius of Argos. Shrub.

RACEMOSA, ras-em-*o*-sa, resembling a raceme.

**Daphne**, *daf*-ne; Gr. name of *Laurus nobilis*, some authorities say named after the river god's daughter in Grecian mythology, who, on being pursued by Apollo, prayed for aid, and was transformed into a Laurel tree. Shrubs.

BLAGAYANA, blag-ay-*a*-na, discovered by Count Blagay.

CNEORUM, ne-*or*-um, old Greek name.

COLLINA, kol-*li*-na, growing on hills.

INDICA, *in*-dik-a, of India.

LAUREOLA, *law*-re-o-la; a little laurel. The Spurge Laurel.

MEZEREUM, me-*ze*-re-um, probably from Persian, *mazaryum*, the Spurge Olive; some authorities believe the word to signify deadly, the berries being poisonous. The Mezereon.

ODORA, od-*o*-ra, sweetly scented.

PONTICA, *pon*-tik-a, Pontic, the shores of the Black Sea.

RUPESTRIS, roo-*pes*-tris, growing on rocks.

**Darlingtonia,** dar-ling-*to*-ne-a; after Dr. Darlington, an American botanist. Hardy herbaceous foliage plant. Insectivorous.

    CALIFORNICA, kal-if-*or*-nik-a, of California.

**Dasylirion,** das-e-*lir*-ion; from Gr. crowded, referring to crowded leaves and flowers. Greenhouse evergreen foliage plants.

    GLAUCOPHYLLUM, glau-kof-*il*-um, glaucous-coloured—the leaves.

**Datura,** da-*tu*-ra; vernacular E. Indian name. Half-hardy annuals and greenhouse perennials and shrubs.

    CERATOCAULA, ser-at-o-*kaw*-la, horn-stalked.

    CORNUCOPIÆ, kor-nu-*ko*-pe-e, horn of plenty.

    FASTUOSA, fas-tu-*o*-sa, stately.

    INERMIS, in-*er*-mis, unarmed, *i.e.,* no spines.

    METELOIDES, me-tel-*oy*-dez, Metel-like.

    SANGUINEA, san-*gwin*-e-a, blood red.

    STRAMONIUM, stra-*mo*-ne-um, old botanic name for the Thorn Apple of medicine. The Thorn Apple.

    SUAVEOLENS, *swa*-ve-ol-enz, sweet smelling.

**Daucus,** *daw*-kus; from the ancient Gr. name.

    CAROTA, ka-*ro*-ta, red rooted. The Carrot.

**Davallia,** da-*val*-le-a; after Edmund Davall, a Swiss botanist. Tropical and temperate ferns.

    BULLATA, bul-*la*-ta, blistered.

    CANARIENSIS, ka-nar-e-*en*-sis, of Canary Islands.

    DECORA, dek-*or*-a, neat or becoming.

    ELEGANS, *el*-e-ganz, elegant.

    FIJIENSIS, fee-jee-*en*-sis, of Fiji Islands.

    MARIESII, mar-*ees*-ei, after Maries, plant collector.

    MOOREANA, moor-e-*an*-a, after Moore.

    TYERMANNII, ty-er-*man*-ei, after Tyermann.

**Davidia,** da-*vid*-e-a; after Armand David, French missionary who botanised in China. Trees.

    INVOLUCRATA, in-vol-u-*kra*-ta, involucred.

**Decumaria,** dek-u-*ma*-re-a; probably from L. *decuma,* a tenth, the calyx and seed-pods often being in tens. Twining plants.

    BARBARA, *bar*-bar-a, foreign.

**Deinanthe,** de-in-*an*-the; from Gr. *deinos* wonderful, and *anthos,* a flower. Herbaceous perennial.

    CŒRULEA, ser-*u*-le-a, sky-blue or azure.

**Delosperma,** de-lo-*sper*-ma; from Gr. *delos,* manifest, and *sperma,* seed. Greenhouse succulents.

    COOPERI, *koo*-per-e, after Cooper.

**Delphinium,** del-*fin*-e-um; from Gr. *delphin,* a dolphin, from the fancied resemblance of the flower-spur to a dolphin's head. Annuals, biennials, and herbaceous perennials.

    AJACIS, aj-*a*-kis; this name has been stated to be founded on some marks at the base of the united petals which were compared to the letters AIAI.

    CARDINALE, kar-din-*a*-le, scarlet.

    CONSOLIDA, kon-*sol*-id-a, joined in one.

    ELATUM, e-*la*-tum, tall.

    FORMOSUM, for-*mo*-sum, beautiful.

    GRANDIFLORUM, gran-dif-*lo*-rum, large flowered.

    NUDICAULE, nu-dik-*aw*-le, naked-stemmed.

    SULPHUREUM, sul-fur-*e*-um, sulphur-yellow.

    TATSIENSE, tat-*se*-en-se, of Tatsien, China.

    TRISTE, *tris*-te, sad, the dull blue of the flowers.

    ZALIL, *zal*-il, after Zalil, native Afghanistan name.

    Most of the delphiniums of gardens are hybrids raised from various species, including elatum, formosum, grandiflorum and sulphureum.

**Dendrobium,** den-*dro*-be-um; from Gr. *dendron,* a tree, and *bios,* life, in allusion to the wild plants being found on trees. Tropical orchids.

    BIGIBBUM, big-*ib*-bum, two-humped.

    CHRYSOTOXUM, kris-ot-*ox*-um, golden arched.

    CHRYSANTHUM, kris-*anth*-um, golden flowered.

    DALHOUSIANUM, dal-*hou*-si-a-num, after Dalhouse.

    DENSIFLORUM, den-sif-*lor*-um, dense flowered.

    FIMBRIATUM, fim-bre-*a*-tum, fringed.

    INFUNDIBULUM, in-fun-*dib*-u-lum, funnel-shaped.

    NOBILE, *no*-bil-e, noble, fine.

    PRIMULINUM, prim-u-*le*-num, primrose coloured.

    SPECIOSUM, spes-e-*o*-sum, showy.

    WARDIANUM, ward-e-*a*-num, after Ward.

**Dendrochilum,** den-dro-*ki*-lum; from Gr. dendron, a tree, and *cheilos*, a lip, referring to lipped flowers and growing on trees. Tropical orchids.

COBBIANUM, kob-i-*a*-num, after Cobb.
FILIFORME, fil-if-*or*-me, thread-like.
GLUMACEUM, glu-*ma*-ce-um, flower spike grass-like.

**Dendromecon,** den-*drom*-e-kon; from Gr. *dendron*, a tree, and *mecon*, a poppy. Shrubby perennial.

RIGIDUM, *rig*-id-um, rigid. The Tree Poppy.

**Dentaria,** den-*tar*-e-a; from L. *dens*, a tooth, in reference to the tooth-like scales on the roots. Herbaceous perennials.

BULBIFERA, bul-*bif*-er-a, bearing bulblets.
ENNEAPHYLLA, en-ne-af-*il*-la, nine-leaved —leafleted.
DIGITATA, dij-it-*a*-ta, five-fingered or lobed—the leaves.

**Desfontainea,** des-fon-*ta*-ne-a; after M. Desfontaines, a French botanist. Shrub.

SPINOSA, spin-*o*-sa, spiny.

**Desmodium,** des-*mo*-de-um; from Gr. *desmon*, chain, referring to pointed pod. Tender shrubs.

CANADENSE, kan-a-*den*-se, of Canada.
GYRANS, *gy*-ranz, gyrating or turning round. The Telegraph Plant.
TILIÆFOLIUM, til-e-e-*fo*-le-um, leaved like Tilia (lime or linden).

**Deutzia,** *doyts*-e-a; named in commemoration of J. van der Deutz, patron of Thunberg. Shrubs.

CORYMBOSA, kor-im-*bo*-sa, flowers in a broad corymb.
CRENATA, kre-*na*-ta, crenate or scalloped leaves.
DISCOLOR, *dis*-ko-lor, various colours— the flowers.
GRACILIS, *gras*-il-is, slender.
SCABRA, *ska*-bra, rough, the bark.
SETCHUENENSIS, setsh-u-en-*en*-sis, of Szechuen.
SIEBOLDIANA, se-bold-e-*an*-a, after P. F. von Siebold, botanist and author.
VILMORINÆ, vil-more-*e*-ne, after Madame de Vilmorin.

**Dianthus,** di-*an*-thus; from Gr. *dios*, a god or divine, and *anthos*, a flower; the Divine flower. Annuals, biennials and perennials. The Carnation and Pink family.

ALLWOODII, all-*wood*-ei, after Montague Allwood, the originator; a race of hybrid pinks.

**Dianthus** (*continued*)

ANATOLICUS, an-at-*ol*-ik-us, of Anatolia.
ATRORUBENS, at-ro-*ru*-bens, dark red.
BARBATUS, bar-*ba*-tus, bearded. The Sweet William.
CÆSIUS, *se*-se-us, grey, or bluish-grey. The Cheddar Pink.
CALLIZONUS, kal-e-*zo*-nus, beautifully zoned.
CARTHUSIANORUM, kar-*thu*-se-an-*or*-um, of the Carthusians.
CARYOPHYLLUS, kar-e-of-*il*-lus, old name of the shrub, the flower buds of which are the cloves of commerce; given here on account of the clove-like fragrance. The Carnation and Clove.
CRUENTUS, kru-*en*-tus, blood-red, the flowers.
DELTOIDES, del-*toy*-des, triangular, shape of Greek letter *delta* on the petals. The Maiden Pink.
GLACIALIS, glas-e-*a*-lis, icy, from snow regions.
GRANITICUS, gran-*it*-ik-us, of granite rocks.
MICROLEPIS, mi-kro-*le*-pis, having small scales.
NEGLECTUS, neg-*lek*-tus, unobserved or unnoticed, the plant being inconspicuous when flowers are closed.
PLUMARIUS, plu-*mar*-e-us, the flowers feathered, or frilled. The Pink.
SQUARROSUS, skwar-*ro*-sus, rough or scurfy.
SINENSIS, sin-*en*-sis, of China. The Chinese or Indian Pink.
SUPERBUS, sup-*er*-bus, superb. The Fringed Pink.

**Diascia,** di-*as*-ke-a; from Gr. words for two and pouch, referring to the two-spurred corolla. Low slender herbs.

BARBERÆ, *bar*-ber-e, after Mrs. Barber.

**Dicentra,** di-*sen*-tra; from Gr. *di*, two, and *kentron*, a spur, alluding to the two spurs of the petals. Hardy herbaceous.

CHRYSANTHA, kris-*anth*-a, golden-flowered.
CUCULLARIA, ku-kul-*lar*-e-a, hooded.
EXIMIA, eks-*im*-e-a, choice.
FORMOSA, for-*mo*-sa, handsome.
SPECTABILE, spek-*tab*-il-e, showy. Lyre Flower or Bleeding Heart.

**Dichorisandra,** dik-or-e-*zan*-dra; from Gr. *dis*, twice, *chorizo*, to part, and *aner*, a man, alluding to the anthers being two-cleft or two-celled. Tropical flowering and foliage plants.

MUSAICA, mu-*sa*-ik-a, mosaic variegation.

**Dichorisandra** (*continued*)

THYRSIFLORA, thir-sif-*lor*-a, thyrse flowered.

**Dicksonia**, dik-*so*-ne-a; after Jas. Dickson, a botanist. Greenhouse tree ferns.

ANTARCTICA, ant-*ark*-tik-a, southern, or opposite the Arctic.

SQUARROSA, skwar-*ro*-sa, rough, shaggy.

**Dictamnus**, dik-*tam*-nus; classical name adopted from Virgil, the plant once being common on Mt. Dicte. Hardy herbaceous.

FRAXINELLA, fraks-in-*el*-la, from *fraxinus*, L. for an ash, the leaves resembling those of that tree.

**Didiscus**, did-*is*-kus; from Gr. *di*, twice, and *diskos*, a flat disk; the form of the flower head. Half-hardy annual.

CÆRULEUS, ser-*u*-le-us, sky-blue.

**Dieffenbachia**, deef-fen-*bak*-e-a; after Dr. Dieffenbach, a German botanist. Tropical foliage plants.

BAUSEI, *baws*-e-i, after Bause.

MEMORIA CORTI, me-*mor*-e-a *kort*-i, in memory of Cort.

SHUTTLEWORTHII, shutl-*worth*-ei, after Shuttleworth.

**Dielytra**, di-el-*it*-ra; from Gr. *di*, two, and *elytræ*, wings; two-winged. Hardy herbaceous flowering plants; this genus is now included in Dictamnus.

**Dierama**, di-er-*a*-ma; from Gr. *dierama*, a funnel, alluding to the shape of the flowers. Bulbous perennial.

PULCHERRIMUM, pul-*ker*-rim-um, most beautiful.

**Diervilla**, di-er-*vil*-la; after M. Dierville, a famous French surgeon. Flowering shrubs.

FLORIBUNDA, flor-ib-*un*-da, abundant flowers.

FLORIDA, *flor*-id-a, richly flowering.

ROSEA, *ro*-ze-a, rose-coloured.

**Digitalis**, dij-it-*a*-lis; from L. *digitus*, a finger, the flower resembling the finger of a glove. Biennials and herbaceous perennials.

FERRUGINEA, fer-ru-*jin*-e-a, rusty—the flowers brownish.

LUTEA, *lu*-te-a, yellow.

PURPUREA, pur-*pur*-e-a, purple. The Foxglove.

**Dimorphanthus**, di-mor-*fan*-thus; Gr. *di*, two, *morphe*, shape, *anthos*, a flower, flowers of two forms. Foliage shrubs.

MANDSCHURICUS, mands-*kur*-ik-us, of Manchuria.

**Dimorphotheca**, di-mor-*foth*-ek-a; from Gr. *di*, two, *morphe*, shape, and *theka*, a case or capsule, in reference to the two formed achenes. Annuals and perennials.

AURANTIACA, aw-ran-te-*a*-ka, golden-orange.

PLUVIALIS, plu-ve-*a*-lis, rainy, presumably alluding to natural season of flowering.

**Dionæa**, di-o-*ne*-a; from Gr. *Dione*, the mother of Venus. Greenhouse herbaceous insectivorous plant.

MUSCIPULA, mus-*kip*-u-la, fly catching. The Venus's Fly Trap.

**Dioscorea**, di-os-*kor*-e-a, after Dioscorides, a Greek physician. Hardy tuberous-rooted climber.

BATATAS, ba-*ta*-tas, native name and reputed origin of our word Potato. The Yam.

**Diosma**, di-*oz*-ma; from Gr. *dios*, divine, and *osme*, odour, referring to the perfume. Greenhouse shrub with fragrant leaves.

CAPITATA, kap-it-*a*-ta, flowers in heads or clusters.

ERICOIDES, er-ik-*oy*-dez, like Erica or heather.

**Dioscorea**, di-os-*kor*-e-a, after Dios-divine, and *puros*, wheat or food—lit. food for the gods. Deciduous fruiting trees. Persimmon or Date Plum.

KAKI, *ka*-ke, native name. The Persimmon of Japan.

LOTUS, *lo*-tus, after Lotus, which if eaten causes forgetfulness. The Persimmon of Southern Europe.

VIRGINIANA, ver-jin-e-*a*-na, of Virginia. The Virginian Date Plum.

**Dipelta**, di-*pel*-ta; from Gr. *di*, two, and *pelta*, a shield, referring to the pair of opposite bracts. Shrubs.

FLORIBUNDA, flor-ib-*un*-da, many-flowered.

VENTRICOSA, ven-trik-*o*-sa, inflated, appearing swollen.

**Diplacus**, dip-la-kus; from Gr. *dis*, two, and *plakos*, placenta, seed capsule with two placentas. Greenhouse flowering shrubs.

**Diplacus** (*continued*)
GLUTINOSUS, glu-tin-*o*-sus, sticky.
PUNICEUS, pu-*nik*- (or *nis*-) e-us, bright red.

**Dipladenia,** dip-la-*de*-ne-a; from Gr. *diploos*, double, and *aden*, a gland; two gland-like developments on the ovary. Tropical evergreen flowering climbers.
AMABILIS, am-*a*-bil-is, lovable or amiable.
BOLIVIENSIS, bol-iv-e-*en*-sis, of Bolivia.
BREARLEYANA, brear-le-*a*-na, after Brearley.

**Diplopappus,** dip-lo-*pap*-pus; from Gr. *diploos*, double, and *pappos*, a plume, in allusion to the feathery awns which crown the seeds. Shrub.
CHRYSOPHYLLUS, kris-of-*il*-lus, golden —the leaves.

**Disa,** *dy*-sa; origin of name unknown. Greenhouse terrestrial orchids.
GRANDIFLORA, gran-dif-*lor*-a, large flowered. Table Mountain Orchis.

**Dipsacus,** *dip*-sa-kus; from Gr. *dipsao*, to thirst, referring to the cavity formed by the uniting of the leaves round the stem, which collects and holds water. Biennials.
FULLONUM, ful-*lo*-num, fullers'. The Fullers' Teasel.
INERMIS, in-*er*-mis, unarmed.
SYLVESTRIS, sil-*ves*-tris, of woodland. The Teasel.

**Disocactus,** dis-o-*kak*-tus; from Gr. *dis*, two or twice and *cactus*, two-shaped cactus. Greenhouse cacti.
BIFORMIS, bif-*or*-mis, two forms—simulating both epiphyllum and phyllocactus.

**Dodecatheon,** do-dek-*ath*-e-on; ancient name signifying "Flower of the Twelve Gods"; from Gr. *dodeka*, twelve, and *theos*, god. Herbaceous perennials.
CLEVELANDI, *kleeve*-land-e, after Cleveland.
HENDERSONI, hen-der-*so*-ne, after Henderson.
JEFFREYI, *jef*-re-e, after John Jeffrey, botanist and collector.
MEADIA, *me*-de-a, after Dr. Mead. Meadia was first suggested as the name for the genus, but not accepted by Linnæus, who chose Dodecatheon.

**Dolichos,** *dol*-ik-os; from old Gr. name of a bean. Greenhouse evergreen twiner.
LABLAB, *lab*-lab, Egyptian name adopted by Linnæus.

**Dolichothele,** dol-ik-o-*the*-le; from Gr. *dolikos*, long, and *thele*, a nipple, the elongated tubercles. Greenhouse cacti.
LONGIMAMMA, long-im-*am*-ma, long tubercled.

**Doodia,** *dood*-e-a; after Samuel Doody, a London cryptogamic botanist. Greenhouse evergreen ferns.
ASPERA, *as*-per-a, rough to touch.
CAUDATA, kau-*da*-ta, tailed—the fronds.

**Doronicum,** dor-*on*-ik-um; from the Arabic name *doronigi* or *doronakh*. Herbaceous perennials.
AUSTRIACUM, *aws*-tre-ak-um, of Austria.
CAUCASICUM, kaw-*kas*-ik-um, of Caucasia.
PLANTAGINEUM, plan-ta-*jin*-e-um, Plantago (plantain)-leaved.

**Douglasia,** dug-*la*-se-a; in commemoration of David Douglas, botanical traveller in N.W. America. Rock plants.
VITALIANA, vit-al-e-*a*-na, Vitalian.

**Downingia,** down-*ing*-e-a; after A. J. Downing, an American patron of botany and landscape gardener. Annuals.
ELEGANS, *el*-e-ganz, elegant.
PULCHELLA, pul-*kel*-la, pretty (diminutive).

**Draba,** *dra*-ba; old Gr. name for a cress, possibly because the leaves of most of the crucifers have a biting taste. Rock plants.
AIZOIDES, ay-*zoy*-des, like aizoon.
AIZOON, ay-*zo*-on, ever-living—the fleshy leaves not fading quickly.
BRUNIÆFOLIA, bru-ne-e-*fo*-le-a, Brunialeaved.
DEDEANA, *ded*-e-an-a, after Dede.
PYRENAICA, pir-en-*a*-ik-a, of Pyrenees.
SUNDERMANI, soon-der-*man*-e, after Herr Sundermann, botanist and collector.

**Dracæna,** drak-*e*-na; from Gr. *drakaina*, a female dragon, juice when thickened supposed to resemble dragon's blood. Name commonly pronounced dra-*see*-na. Tropical foliage plants; many garden hybrids.
AMABILIS, am-*a*-bil-is, amiable or lovable.
BAUSEI, *baus*-e-e, after Bause.
DRACO, *drak*-o, dragon.
GODSEFFIANA, god-*sef*-e-a-na, after Joseph Godseff.
GOLDIEANA, gold-e-*a*-na, after Goldie.
TERMINALIS, ter-min-*a*-lis, terminal or ending—branches ending in inflorescence.

**Dracocephalum**, drak-o-*sef*-a-lum; from Gr. *drakon*, a dragon, and *kephale*, a head, in allusion to the gaping mouth of the flower. Annuals and perennials. Dragon's Head.

FORRESTII, *for*-res-tei, after George Forrest, plant collector.
MOLDAVICA, mol-*da*-vik-a, Moldavian.
NUTANS, *nu*-tans, nodding.
RUYSCHIANUM, rews-ki-*an*-um, after Ruysch.
VIRGINIANA, vir-jin-e-a-na, of Virginia.

**Dracunculus**, drak-*un*-ku-lus, an ancient name and diminutive of *draco*, dragon. Hardy tuberous perennial. The Dragon Arum.

VULGARIS, vul-*gar*-is, common.

**Drimys**, *drim*-is; from Gr. *drimus*, acid, referring to the pungent taste of bark and foliage. Half-hardy shrubs.

WINTERI, *win*-ter-i, after Winter, the aromatic bark (Winter's Bark) having been introduced by a Capt. Winter, of one of Drake's ships.

**Drosera**, *dros*-er-a; from Gr. *droseros*, dewy, referring to the dew-like glands on the leaves. Hardy and greenhouse bog plants; insectivorous.

CAPENSIS, ka-*pen*-sis, of the Cape.
DICHOTOMA, dik-*ot*-o-ma, twiced forked.
LONGIFOLIA, long-if-*o*-le-a, long-leaved.
ROTUNDIFOLIA, rot-un-dif-*o*-le-a, round-leaved. The Sundew.

**Dryas**, *dri*-as; from Gr. *druas*, a dryad, or wood-nymph; some authorities derive from Gr. *drus*, an oak, leaves of plant resemble those of oak. Rock plants.

DRUMMONDII, drum-*mon*-dei, after Drummond.
OCTOPETALA, ok-to-*pet*-a-la, eight-petalled.

**Eccremocarpus**, ek-krem-o-*kar*-pus;from Gr. *ekkremes*, pendant, and *karpos*, fruit, the seed vessels being pendulous. Half-hardy, sub-shrubby climber.

SCABER, *ska*-ber, rough—to the touch.

**Echeveria**, esh-e-*ve*-re-a, or ek-e-*veer*-e-a; after A. Echeverria, Mexican botanical draughtsman. Tender succulents, some used for summer bedding.

METALLICA, me-*tal*-lik-a, foliage of metallic colouring.
RETUSA, ret-*u*-sa, blunt leaved.
SECUNDA GLAUCA, sek-*un*-da *glaw*-ka, one-sided flower spike and glaucous leaves.

**Echinacea**, ek-in-*a*-se-a; from Gr. *echinos*, hedgehog, referring to the prickly involucre. Hardy herbaceous plants.

PURPUREA, pur-*pur*-e-a, purple coloured.

**Echinocactus**, *ek*-in-o-*kak*-tus; from Gr. *echinos*, a hedgehog, and *cactus*, referring to the spiny plants. Greenhouse cacti.

GRUSONII, *grus*-o-nei, after Gruson.
HORIZONTHALONIUS, hor-e-zon-thal-*on*-e-us, with level areoles.
INGENS, *in*-genz, huge.
PLATYCANTHUS, plat-ik-*an*-thus, broad spined.
POLYCEPHALUS, pol-ik-*ef*-a-lus, many headed.

**Echinocereus**, ek-in-o-*se*-re-us; from Gr. *echinos*, a hedgehog, and *cereus*, an allusion to the prickly plants. Greenhouse cacti.

CHLORANTHUS, klo-*ranth*-us, green—the flowers.
DASYACANTHUS, das-e-ak-*an*-thus, hairy flowered.
FENDLERI, *fend*-ler-i, after Fendler.
PECTINATUS, pek-tin-*a*-tus, like a comb, the spines.
PULCHELLUS, pul-*kel*-lus, small and beautiful.
VIRIDIFLORUS, vir-id-if-*lor*-us, green flowered.

**Echinofossulocactus**, *ek*-in-o-*fos*-sul-o-*kak*-tus, from Gr. *echinos*, a hedgehog, and L. *fossula*, a groove, and *cactus*—the plants prickly and grooved.

COPTONOGONUS, kop-ton-*og*-on-us, notched or wavy—the ribs.
CRISPATUS, kris-*pa*-tus, curled.
OBVALLATUS, ob-val-*la*-tus, fortified as with a rampart.

**Echinops**, *ek*-in-ops, from Gr. *echinos*, a hedgehog, and *opsis*, like, in reference to the spines which surround the flowers. Herbaceous perennials and biennials.

BANATICUS, ban-*nat*-ik-us, from Banat.
RITRO, *ri*-tro, S. European name.
SPHÆROCEPHALUS, sfer-o-*sef*-a-lus, round headed.

**Echinopsis**, ek-in-*op*-sis; from Gr. *echinos*, a hedgehog, and *opsis*, like—the plants are balls of spines. Greenhouse cacti.

EYRIESII, ey-*ree*-sei, after Eyries.
OXYGONUS, ox-ig-*o*-nus, sharp-angled.

**Echium**, *ek*-e-um; from Gr. *echis*, a viper, the seeds being supposed to resemble a viper's head; some authorities contend that the word alludes to

**Echium** (*continued*)
the old belief in the plant (Viper's Bugloss) as a remedy for adder's bite. Or from Gr. *echion*, the ancient name for this plant. Annuals, biennials, and perennials.

ALBICANS, *al*-bik-ans, whitish.

CRETICUM, *kret*-ik-um, Cretan.

PLANTAGINEUM, plan-ta-*jin*-e-um, plantain (Plantago)-like.

**Edwardsia**, ed-*ward*-se-a; named after Sydenham Edwards, a botanical artist. Shrubs.

GRANDIFLORA, gran-dif-*lo*-ra, large flowered.

**Eichhornia**, ish-*hor*-ne-a; after J. A. F. Eichhorn, a Prussian. Floating aquatics. The Water Hyacinth.

AZUREA, az-*u*-re-a, blue.

CRASSIPES, *kras*-sip-eez, thick stalked.

SPECIOSA, spes-e-*o*-sa, showy.

**Elæagnus**, el-e-*ag*-nus, from Gr. *elaia*, an olive tree, and *agnos*, a plant like a willow; or from *elæagnus*, the name given by Dioscorides, physician of ancient Greece, to the Wild Olive. Shrubs.

ARGENTEA, ar-*jen*-te-a, silvery—the leaves.

GLABRA, *gla*-bra, smooth.

MACROPHYLLA, mak-ro-*fil*-la, large-leaved.

MULTIFLORA, mul-tif-*lo*-ra, many-flowered.

PUNGENS, *pun*-jens, piercing—the twigs are thorny.

**Elisena**, el-is-*e*-na; origin unknown. Greenhouse bulbous perennial.

LONGIPETALA, long-ip-*et*-a-la, long petalled.

**Elatine**, e-*lat*-e-ne; Gr. name of doubtful application. Submerged aquatics.

HEXANDRA, hex-*an*-dra, six-stamened.

MACROPODA, mak-ro-*po*-da, large rooted —forms big masses.

**Elodea**, el-*o*-de-a; from Gr. *elodes*, a marsh—the native habitat. Submerged aquatics.

CANADENSIS, kan-a-*den*-sis, of Canada. The Water Thyme.

DENSA CRISPA, *den*-sa *kris*-pa, foliage tightly curled.

**Embothrium**, em-*both*-re-um; from Gr. name referring to the structure of the anthers. Shrubs.

COCCINEUM, kok-*sin*-e-um, scarlet flowers.

**Emilia**, em-*il*-e-a; derivation unknown, probably commemorative. Half-hardy annual.

SAGITTATA, sag-it-*ta*-ta, arrow-head-like —the leaves.

**Empetrum**, em-*pe*-trum, from Gr. *en*, in, and *petros*, a rock, the plant haunting rocky places. Shrub.

NIGRUM, *nig*-rum, black, the berries.

**Enkianthus**, en-ki-*an*-thus; from Gr. *enkuos*, enlarged, and *anthos*, a flower, presumably in reference to the round, bell-shaped blossoms. Shrubs.

CAMPANULATUS, kam-pan-u-*la*-tus, bell-shaped.

CERNUUS, *ser*-nu-us, drooping—the flowers.

JAPONICUS, jap-*on*-ik-us, of Japan.

**Eomecon**, e-om-*e*-kon; from Gr. *eos*, the dawn, and *mecon*, a poppy, the Japanese "Poppy of the Dawn." Hardy perennial.

CHIONANTHA, ki-on-*an*-tha, snowy—the white flowers.

**Epacris**, e-*pak*-ris, or *ep*-ak-ris; from Gr. *epi*, upon, and *akros*, the top; habitat on high ground. Greenhouse flowering shrubs.

HYACINTHIFLORA, hy-a-*sinth*-if-*lo*-ra, hyacinth-flowered.

IMPRESSA, im-*pres*-sa, flower tube with two depressions.

LONGIFLORA, long-if-*lo*-ra, long-flowered.

**Epidendrum**, ep-i-*den*-drum; from Gr. *epi*, upon, and *dendron*, a tree—growing upon trees. Epiphytal orchids.

FRAGRANS, *fra*-granz, fragrant.

RADICANS, *rad*-e-kans, producing roots.

VITELLINUM, vit-el-*le*-num, yolk-of-egg colour.

WALLISII, wal-*lis*-ci, after Wallis.

**Epigæa**, ep-e-*je*-a; from Gr. *epi*, upon, and *ge*, the earth, in reference to the plant's prostrate habit. Creeping shrub.

REPENS, *re*-penz, creeping and rooting.

**Epilobium**, ep-e-*lo*-be-um; from Gr. *epi*, upon, and *lobos*, a pod, the flowers appearing to be growing on the seed-pod. Herbaceous perennials. The Willow Herbs.

ANGUSTIFOLIUM, an-gus-tif-*o*-le-um, narrow-leaved. The Rose Bay.

DODONÆI, dod-on-*a*-e, after Dodon.

FLEISCHERI, *fli*-sher-i, after Fleischer.

GLABELLUM, glab-*el*-um, small and smooth—the leaves.

**Epilobium** (*continued*)

HIRSUTUM, hir-*su*-tum, hairy. The Codlins and Cream.

NUMMULARIÆFOLIUM, num-mul-*ar*-e-e-*fo*-le-um, rounded leaves simulating money.

OBCORDATUM, ob-kor-*da*-tum, obversely heart-shaped.

ROSMARINIFOLIUM,rose-mar-in-if-*o*-le-um, Rosemarinum (rosemary)-leaved.

**Epimedium**, ep-im-*e*-di-um; Gr. name of obscure meaning. Herbaceous and rock plants. The Barrenworts.

ALPINUM, al-*pine*-um, alpine.

MACRANTHUM, mak-*ranth*-um, large-flowered.

PINNATUM, pin-*na*-tum, pinnate-leaved.

**Epipactis**, ep-ip-*ak*-tis; from Gr. *epipegnuo*, to coagulate—its effect on milk. Terrestrial orchids.

LATIFOLIA, la-tif-*o*-le-a, broad-leaved.

PALUSTRIS, pal-*us*-tris, found in marshy places.

**Epiphyllum**, ep-e-*fil*-lum; from Gr. *epi*, upon, and *phyllon*, a leaf—flowers borne on leaf-like branches. Greenhouse cacti.

GÆRTNERI, geert-ner-i, after Gærtner.

RUSSELLIANUM, rus-sel-e-*a*-num, after Russell.

TRUNCATUM, *trun*-ka-tum, blunt, as if cut off.

**Epithelantha**, ep-e-thel-*an*-tha; from Gr. *epi*, upon, *thele*, a nipple, and *anthos*, a flower, the flowers are borne upon the tubercles. Greenhouse cacti.

MICROMERIS, mi-kro-*mer*-is, small in all its parts.

**Equisetum**, ek-we-*se*-tum; from L. *equus*, horse, and *seta*, a bristle, the barren growths resembling horses' tails. Herbaceous perennial. The Horsetails.

SYLVATICUM, sil-*vat*-ik-um, woodland habitat.

**Eragrostis**, er-a-*gros*-tis, from *eros*, love, and *agrostis*, grass, in allusion to its beauty. Ornamental annual grass.

ELEGANS, *el*-e-ganz, elegant. The Love Grass.

**Eranthemum**, e-*ran*-them-um; from Gr. *erao*, to love, and *anthos*, a flower, referring to the beauty of the flowers. Tropical sub-shrubby flowering plants.

ALBIFLORUM, al-bif-*lor*-um, white-flowered.

ANDERSONII, an-der-*so*-nei, after Anderson.

COOPERI, *koo*-per-i, after Cooper.

**Eranthis**, e-*ran*-this; from Gr. *er*, spring, and *anthos*, a flower, referring to the early flowering. Tuberous perennials.

CILICICA, si-*lis*-ik-a, Cilician.

HYEMALIS, hi-em-*a*-lis, pertaining to winter. The Winter Aconite.

**Ercilla**, er-*sil*-la; Peruvian name. Climbing evergreen shrubs native of Peru and Chile. Hardy evergreen creeper.

VOLUBILIS, vol-*u*-bil-is, twisting round.

**Eremurus**, er-e-*mu*-rus; from Gr. *eremos*, solitary, and *ioura*, a tail, alluding to the single flower-spike. Herbaceous perennials. The Fox-tail Lily.

BUNGEI, *bun*-je-e, after Bunge, a Russian botanist.

ELWESI, el-*we*-se, after H. J. Elwes, botanical author and traveller.

HIMALAICUS, him-le-*a*-ik-us, Himalayan.

ROBUSTUS, ro-*bus*-tus, robust.

**Erica**, er-ik-a (more correctly e-*ri*-ka); from Gr. *ereike*, heath or heather. Shrubs.

ARBOREA, ar-*bor*-e-a, tree-like.

AUSTRALIS, aws-*tra*-lis, southern.

CARNEA, *kar*-ne-a, flesh-coloured.

CAVENDISHIANA, kav-en-dish-e-*a*-na, after Cavendish.

CILIARIS, sil-e-*ar*-is, fringed with fine hairs—the leaves.

CINEREA, sin-er-*e*-a, grey or ashen—under parts of leaves.

DARLEYENSIS, dar-le-*en*-sis, originated at a Darley Dale nursery.

GRACILIS, *gras*-il-is, slender.

HYEMALIS, hi-em-*a*-lis, pertaining to winter.

LUSITANICA, lu-sit-*a*-nik-a, of Lusitania (Portugal).

MEDITERRANEA, med-it-er-*ra*-ne-a, of Mediterranean region.

MELANTHERA, mel-*an*-ther-a, black anthers.

STRICTA, *strik*-ta, upright.

TETRALIX, te-*tra*-liks, four leaves arranged crosswise.

VAGANS, *va*-gans, wandering, widespread.

VENTRICOSA, ven-trik-*o*-sa, swollen or bellied.

**Erigeron**, er-*ij*-er-on; from Gr. *eri*, early (or *ear*, spring), and *geron*, old (an old man), alluding to the hairy seed pappus; or, more probably, to the hoary appearance of the leaves of some species in spring. Herbaceous perennials.

COULTERI, *kol*-ter-e, after Coulter.

MUCRONATUS, mu-kron-*a*-tus, small-pointed—the leaves.

**Erigeron** (*continued*)

MULTIRADIATUS, mul - te - *ra* - de - a - tus, many-rayed.

PHILADELPHICUS, fil-a-*del*-fik-us, of Philadelphia, U.S.A.

SPECIOSUS, spes-e-*o*-sus, handsome.

STRIGOSUS, strig-*o*-sus, bristly.

**Erinacea**, er-in-*a*-se-a; from L. *erinaceus*, a hedgehog, the branches being spiny; some authorities give Gr. *erineos*, woollen, the calyx being very hairy. Shrubs.

PUNGENS, *pun*-jens, spiny.

**Erinus**, er-*i*-nus; from Gr. *eri*, early, the flowers appearing in spring. Rock plants.

ALPINUS, al-*pine*-us, alpine.

**Eriobotrya**, er-e-o-*bot*-re-a; from Gr. *erion*, wool, and *botrys*, a bunch of grapes, reference to the downy flower clusters. Half-hardy fruiting evergreen shrub.

JAPONICA, jap-*on*-ik-a, of Japan. The Loquat.

**Eriocaulon**, er-e-o-*kau*-lon; from Gr. *erion*, wool, and *kaulon*, a stem, alluding to the woolly stems of some species. Submerged aquatics.

SEPTANGULARE, sep-tang-ul-*a*-re, seven angled—the flower scapes. The Pipewort.

**Eriogonum**, er-e-o-*go*-num; from Gr. *erion*, wool, and *gonu*, a joint, the joints of the stems being downy. Hardy perennials.

COMPOSITUM, kom-*pos*-it-um, compound —the flowers in a composite head.

UMBELLATUM, um-bel-*la*-tum, flowers in umbels.

**Eriophorum**, er-e-*of*-or-um; from Gr. *erion*, wool, and *phoreo*, to bear, in reference to the silky flower heads. Aquatic perennials. The Cotton Grass.

POLYSTACHYON, pol-is-*tak*-e-on, many-spiked.

VAGINATUM, va-jin-*a*-tum, sheathing.

**Eriostemon**, er-e-os-*te*-mon, from Gr. *erion*, wool, and *stemon*, a stamen, the downy stamens. Greenhouse shrubs.

BUXIFOLIUM, bux-if-*o*-le-um, box (Buxus)-leaved.

PULCHELLUS, pul-*kel*-lus, beautiful.

**Erodium**, er-*o*-deum; from Gr. *erodios*, a heron, in reference to the resemblance of the style and ovaries to the head and beak of that bird. Herbaceous and rock plants. The Heron's-bill.

**Erodium** (*continued*)

AMANUM, am-*a*-num, from Mt. Amano, Syria.

CHAMÆDRYOIDES, kam-e-dri-*oy*-des, like chamædrys (Germander).

CHRYSANTHUM, kris-*anth*-um, golden-flowered.

CORSICUM, *kor*-sik-um, of Corsica.

GUTTATUM, gut-*ta*-tum, spotted—the flowers.

MACRADENUM, mak-ra-*de*-num, large-glanded—the foliage.

MANESCAVII, man-es-*ka*-vei, after Manescave.

OLYMPICUM, ol-*im*-pe-kum, Olympian, *i.e.*, Mt. Olympus.

PELARGONIFOLIUM, pel-ar-go-nif-*o*-le-um, pelargonium-leaved.

TRICHOMANEFOLIUM, trik-o-ma-ne-*fo*-le-um, leaved like Trichomanes, a genus of ferns.

**Eryngium**, e-*ring*-e-um (or er-*in*-je-um); from Gr. name for some sort of thistle with spiny-toothed leaves. Herbaceous perennials. The Eryngoes.

AMETHYSTINUM, am-e-this-*te*-num, violet-blue.

BOURGATII, boor-*ga*-tei, after Bourgati.

MARITIMUM, mar-*it*-im-um, belonging to sea coasts. The Sea Holly.

PLANUM, *pla*-num, flat-leaved.

**Erysimum**, er-*is*-im-um; from Gr. *eruo*, to draw up, some of the species being supposed to raise blisters. Biennial and perennial, rock and border plants.

ARKANSANUM, ar-kan-*sa*-num, of Arkansas, U.S.A.

LINIFOLIUM, li-nif-*o*-le-um, flax (Linum)-leaved.

OCHROLEUCUM, ok-ro-*lew*-kum, yellow-white, or cream-coloured.

PEROFSKIANUM, per-of-*ske*-a-num, after M. Perofsky, a Russian botanist. Also spelled Perowskianum.

**Erythræa**, er-ith-*re*-a, from Gr. *erythros*, red, the colour of the flowers. Annuals and perennials.

CENTAURIUM, sen- (or ken-) *taw*-re-um, old name after the Centaur Chiron, famed in medicine. The Centaury.

MASSONI, *mas*-son-e, after Masson, plant collector.

**Erythrina**, er-ith-*re*-na; from Gr. *erythros*, red, referring to colour of flowers. Half-hardy flowering shrub. The Coral-tree.

CRISTA-GALLI, *kris*-ta-*gal*-li, a cock's comb.

**Erythronium,** er-ith-*ro*-ne-um; from Gr. *erythros*, red, that being the colour of the flowers of earlier species introduced. Bulbous plants.

AMERICANUM, a-mer-ik-*a*-num, American.

CITRINUM, sit-*ri*-num, citron-yellow—the flowers.

DENS-CANIS, dens-*ka*-nis, a dog's-tooth. The Dog's-tooth Violet.

HARTWEGII, hart-*ve*-gei, of Hartweg, German botanist.

HOWELLII, *how*-el-ei, after Howell.

REVOLUTUM, rev-ol-*u*-tum, revolute or rolled back—the leaves.

**Escallonia,** es-kal-*lo*-ne-a, in commemoration of Señor Escallon, a Spanish traveller. Shrubs.

EDINENSIS, e-din-*en*-sis, of Edinburgh.

EXONIENSIS, ex-o-ne-*en*-sis, of Exeter.

FLORIBUNDA, flor-ib-*un*-da, many-flowered.

ILLINITA, il-*lin*-it-a, varnished or glossy —the leaves.

MACRANTHA, mak-*ran*-tha, long-flowered.

MONTEVIDENSIS, *mon*-te-vid-*en*-sis, of Montevideo.

PTEROCLADON, ter-*ok*-la-don, winged branches.

PULVERULENTA, pul-ver-ul-*en*-ta, powdered.

PUNCTATA, pungk-*ta*-ta, speckled or dotted—leaf-undersides.

RUBRA, *roo*-bra, red—the flowers.

**Eschscholtzia,** esh-*sholt*-se-a; after Dr. J. F. von Eschscholtz, naturalist and physician, attached to a Russian exploring expedition to N.W. America a century ago. Usually treated as annuals.

CALIFORNICA, kal-if-*or*-nik-a, of California. The Californian Poppy.

**Escobaria,** es-ko-*bar*-e-a; after two Mexicans, brothers, Romula and Numa Escobar, of Mexico City. Greenhouse cacti.

DASYACANTHA, das-e-a-*kanth*-a, thick spines or spines close together.

TUBERCULOSA, *tu*-ber-ku-*lo*-sa, tubercled.

**Eucalyptus,** u-kal-*ip*-tus; from Gr. *eu*, good or well, and *calypha*, covered, in allusion to the calyx, which covers the flower like a lid. Half-hardy trees and greenhouse foliage plants.

CITRIODORA, sit-re-od-*o*-ra, citron- or lemon-scented—the leaves.

COCCIFERA, kok-*sif*-er-a, bearing scarlet berries.

GLOBULUS, *glob*-ul-us, globular.

**Eucharidium,** u-kar-*id*-e-um; from Gr. *eucharis*, charming, referring to the beauty of these annuals.

BREWERI, *broo*-er-i, after Samuel Brewer, botanist.

CONCINNUM, kon-*sin*-num, neat, pretty.

GRANDIFLORUM, gran-dif-*lo*-rum, large flowered.

**Eucharis,** *u*-kar-is; from Gr. *eucharis*, agreeable, referring to the pleasing fragrance of the flowers. Tropical evergreen bulbous plants.

AMAZONICA, am-a-*zon*-ik-a, Amazonian.

CANDIDA, *kan*-did-a, white.

GRANDIFLORUM, gran-dif-*lo*-rum, large flowered.

**Eucomis,** *u*-kom-is; from Gr. *eukomus*, beautiful haired, referring to the leafy tuft surmounting the flower spike. Greenhouse bulbs.

PUNCTATA, punk-*ta*-ta, dotted—the flower stem.

**Eucryphia,** u-*krif*-e-a; from Gr. *eu*, good or well, and *kryphia*, a covering, in reference to the cap-like cover formed by the calyx. Flowering shrubs.

CORDIFOLIA, kor-dif-*o*-le-a, heart-shaped —the leaves.

PINNATIFOLIA, pin-nat-if-*o*-le-a, pinnate-leaved.

**Eulalia,** u-*lal*-e-a; from Gr. *eu*, well, and *lalia*, speech, referring to the praise bestowed on the genus. Hardy ornamental grasses.

GRACILLIMA, gras-*il*-lim-a, very slender.

JAPONICA, jap-*on*-ik-a, of Japan.

ZEBRINA, ze-*bry*-na, zebra-striped.

**Euonymus,** u-*o*-ne-mus; said to be named after Euonyme, the mother of the Furies in Greek mythology, or from Gr. *euonymos*, of good fame or lucky. (Gr. *eu*, good, and *onoma*, a name, *i.e.*, of good repute.) Often pronounced u-*on*-e-mus. Shrubs.

ALATUS, al-*a*-tus, winged—the branches.

ATROPURPUREUS, *at*-ro-pur-*pur*-e-us, dark purple.

EUROPÆUS, u-ro-*pe*-us, European.

JAPONICUS, jap-*on*-ik-us, of Japan.

LATIFOLIUS, lat-if-*o*-le-us, broad-leaved.

RADICANS, *rad*-e-kanz, rooting as it creeps.

**Eupatorium,** u-pat-*or*-e-um; commemorating Mithridates Eupator, King of Pontus, who discovered in one of the species an antidote for poison. Hardy herbaceous plants and shrubs.

**Eupatorium** (*continued*)

AGERATOIDES, aj-er-at-*oy*-dez, like ageratum.

AROMATICUM, ar-o-*mat*-ik-um, aromatic.

CANNABINUM, kan-nab-*ee*-num, like hemp (Cannabis).

IANTHINUM, ee-*anth*-in-um, violet coloured.

PURPUREUM, pur-*pur*-e-um, purple coloured.

RIPARIUM, re-*pair*-e-um, of river banks.

WEINMANNIANUM, vyn-man-e-*a*-num, after Weinmann, a German botanist.

**Euphorbia,** u-*for*-be-a; after Euphorbus, physician to Juba, King of Mauritania. Annuals, perennials, and succulents.

CAPITATA, kap-it-*a*-ta, flowers in a head.

CYPARISSIAS, si-par-*is*-se-as, resembling a cypress—the foliage. The Cypress Spurge.

EPITHYMOIDES, ep-ith-e-*moy*-des, epithymum like.

HETEROPHYLLA, het-er-*of*-il-la, various leaved.

FULGENS, ful-jenz, gleaming.

JACQUINÆFLORA, jak-kwin-e-*flo*-ra, jacquinia-like flowers.

LATHYRIS, *lath*-e-ris, old name for Caper Spurge.

MARGINATA, mar-jin-*a*-ta, margined with another colour—white.

MYRSINITES, mir-sin-*i*-tes, from ancient Greek name signifying myrtle, ref. obscure.

OBESA, o-*be*-sa, obese, or tub-like.

PILOSA, pil-*o*-sa, softly hairy.

POLYCHROMA, pol-e-*kro*-ma, many-coloured.

PULCHERRIMA, pul-*ker*-rim-a, most beautiful. Poinsettia.

SPLENDENS, *splen*-denz, splendid.

WULFENII, wul-*fe*-nei, after Wulfen, a professor of botany.

**Exacum,** *eks*-ak-um; classical name. Tropical flowering plants.

AFFINE, af-*fee*-ne, related.

ZEYLANICUM, ze-*lan*-ik-um, Cingalese.

**Exochorda,** eks-o-*kor*-da; from Gr. *exo*, outside, or external, and *chorde*, a cord or thong, in reference to the structure of the fruits. Shrubs.

ALBERTII, al-*bert*-ei, discovered by Albert Regel.

GIRALDII, jir-*al*-dei, introduced by Père Giraldi.

GRANDIFLORA, gran-dif-*lo*-ra, large-flowered.

RACEMOSA, ras-em-*o*-sa, flowers in racemes.

**Faba,** *fa*-ba; classical name for bean, now included in the genus Vicia.

**Fabiana,** fab-e-*a*-na; in commemoration of F. Fabiano, a Spaniard. Half-hardy shrub.

IMBRICATA, im-bre-*ka*-ta, overlapping leaves.

**Fagus,** *fa*-gus; the Latin name for a Beech tree, some authorities deriving the word from Gr. *phago*, to eat, the seeds being edible. Trees.

SYLVATICA, sil-*vat*-ik-a, inhabiting woods —the Beech Tree.

**Fatsia,** *fat*-se-a; from the Japanese name, *Fatsi*, for F. japonica. Shrubs.

JAPONICA, jap-*on*-ik-a, of Japan.

PAPYRIFERA, pap-e-*rif*-er-a, papery.

**Faucaria,** fau-*kar*-e-a; possibly from L. *fauces*, the throat, the pairs of leaves simulating the open mouths (throats) of animals. Greenhouse succulents.

BOSSCHEANA, *bos*-she-*a*-na, after Bossche.

FELINA, fe-*le*-na, cat-like.

LUPINA, lu-*pe*-na, wolf-like.

TIGRINA, tig-*re*-na, tiger striped.

**Felicia,** fe-*lis*-e-a; named after Herr Felix, a German official. Annuals and perennials.

ABYSSINICA, ab-is-*sin*-ik-a, Abyssinian.

AMELLOIDES, a-mel-*loy*-dez, Aster amellus-like.

FRAGILIS, *fraj*-il-is, slender, fragile.

**Fendlera,** fend-*le*-ra; after Fendler, a plant collector. Shrubs.

RUPICOLA, roo-*pik*-o-la, rock-inhabiting.

**Ferocactus,** fer-o-*kak*-tus; from L. *ferox*, fierce, and *cactus*, refers to the very spiny plants. Greenhouse cacti.

CORNIGERUS, cor-nig-*er*-us, horn bearing.

ECHIDNE, ek-*id*-ne, viper's fang-like— the spines.

LATISPINUS, lat-is-*pin*-us, broad-spined.

LE CONTEI, lec-*on*-te-i, after Le Conte.

UNCINATUS, un-sin-*a*-tus, hooked—the spines.

VIRIDESCENS, vir-id-*es*-senz, greenish flowered.

WISLIZENII, wis-liz-*ee*-nei, after Wislizen.

**Ferula,** *fer*-u-la; name given to the Giant Fennel by Pliny, the Roman naturalist. Herb and perennials.

COMMUNIS, kom-*mu*-nis, gregarious.

FŒTIDA, *fet*-id-a, fetid.

TINGITANA, ting-e-*ta*-na, of Tangiers.

**Festuca**, fes-*tu*-ka; from L. *festuca*, a stem or blade. Grasses.
> OVINA, o-*ve*-na, pertaining to sheep (fodder), hence Sheep's Fescue, var. of glauca, blue-leaved, grown in gardens.

**Ficus**, *fi*-kus; the Latin name for a Fig tree and one common to most European languages. Believed to be derived from the Hebrew name, *fag*.
> CARICA, *kar*-ik-a, of Caria, Asia Minor (ancient geography). The Fig.
> ELASTICA, el-*as*-tik-a, elastic. The India-Rubber Tree.
> PARCELLII, *par*-sel-lei, after Parcell.
> REPENS, *re*-penz, creeping.
> STIPULATA, stip-ul-*a*-ta, bearing stipules.

**Fittonia**, fit-*to*-ne-a; commemorative of Elisabeth and Sarah Mary Fitton, botanical authors. Tropical trailing foliage plants.
> ARGYRONEURA, 'ar-ger-on-*eu*-ra, white veined—the leaves.
> GIGANTEA, ji-*gan*-te-a, of large size.
> PEARCEI, *peers*-ei, after Pearce.

**Fœniculum**, fee-*nik*-u-lum; from L. for hay because of its odour. Perennial herb.
> VULGARE, vul-*gar*-e, common. Fennel.

**Fontinalis**, *fon*-tin-a-lis; from L. *fons*, a fountain—alluding to habitat. Submerged aquatic moss.
> ANTIPYRATICA, an-tip-e-*rat*-ik-a, anti-fire, the plant being used to plug spaces round chimneys in wood houses as a precaution against fire.

**Forsythia**, for-*si*-the-a; after Wm. Forsyth, superintendent of the Royal Gardens, Kensington (1737-1805). Shrubs.
> DENSIFLORA, den-sif-*lo*-ra, many or dense flowered.
> INTERMEDIA, in-ter-*me*-de-a; intermediate *i.e.*, between its two parents (believed to be a hybrid).
> SPECTABILIS, spek-*tab*-il-is, showy.
> SUSPENSA, sus-*pen*-sa, hanging down, the flowers.
> VIRIDISSIMA, vir-id-*is*-sim-a, greenest.

**Fothergilla**, foth-er-*gil*-a; commemorating Dr. John Fothergill, who, over a century ago, made a notable collection of American plants in his Essex garden. Shrubs.
> GARDENII, gar-*de*-nei, of Dr. Garden (U.S.A.), who discovered it.
> MONTICOLA, mon-*tik*-o-la, a mountain dweller.

**Fragaria**, fraj-*ar*-e-a (or fra-*gar*-e-a); from the Latin word for strawberry, *fraga*; or from L. *fragans*, sweet-smelling, alluding to the fruit. Rock plants and fruits. The Strawberry.
> ELATIOR, el-*a*-te-or, higher. The Hautbois Strawberry.
> INDICA, *in*-dik-a, of India.
> MONOPHYLLA, mon-of-*il*-la, single leaved.
> VESCA, *ves*-ka, small or feeble. The Wild Strawberry.

**Franciscea**, fran-*sis*-se-a; after Francis, Emperor of Austria and patron of botany. Tropical flowering shrub.
> CALYCINA, kal-ik-*ee*-na, having a cup-like calyx.

**Francoa**, fran-*ko*-a; after F. Franco, a Spaniard. Greenhouse herbaceous.
> APPENDICULATA, ap-pen-dik-ul-*a*-ta, elongated or lengthened, the flowering stalks; lit. appendaged.
> RAMOSA, ra-*mo*-sa, branched.

**Frankenia**, fran-*ke*-ne-a; after J. Franke, a professor of botany at Upsala. Rock plants.
> LÆVIS, *le*-vis, smooth.
> PULVERULENTA, pul-ver-ul-*en*-ta, powdery.

**Fraxinus**, *fraks*-in-us; Latin name for an Ash tree, probably from Gr. *phrasso*, to fence, the wood being useful for fence-making. Trees.
> EXCELSIOR, ek-*sel*-se-or, taller. The Ash.
> MANDSCHURICA, mands-*hoo*-rik-a, of Manchuria.
> MARIESII, mar-*ees*-ei, after Maries, plant collector.
> OREGONA, or-e-*go*-na, of Oregon.
> ORNUS, *or*-nus, old L. name for Ash tree. The Flowering Ash.

**Freesia**, *free*-se-a; after Dr Freece, a native of Kiel. Very fragrant greenhouse bulbs.
> LEICHTLINII, lecht-*le*-nei, of Leichtlin, a botanist of Austria.
> REFRACTA, re-*frak*-ta, bent back; r. alba, the white form.

**Fremontia**, fre-*mon*-te-a; named after Colonel Fremont, explorer of the West. Shrubs.
> CALIFORNICA, kal-if-*or*-nik-a, of California.
> MEXICANA, meks-e-*ka*-na, Mexican.

**Fritillaria**, frit-il-*lar*-e-a; from L. *fritillus*, a dice-box, the markings on the flower resembling those on a chequer or

**Fritillaria** (*continued*)
chess-board, which is often associated with games of dice. Bulbous plants.

ARMENA, ar-*me*-na, Armenian.

AUREA, aw-*re*-a, golden.

CITRINA, sit-*ri*-na, lemon-yellow.

IMPERIALIS, im-peer-e-*a*-lis, imperial, majestic. The Crown Imperial.

MELEAGRIS, me-le-*a*-gris, Gr. name for guinea-fowl, lit. speckled—the flowers chequered.

**Fuchsia,** *few*-che-a; after Leonard Fuchs, a German botanist of fourteenth century. Shrubs.

CORYMBIFLORA, kor-im-bif-*lo*-ra, cluster flowered.

FULGENS, *ful*-jens, glowing.

GRACILIS, *gras*-il-is, slender.

MACROSTEMMA, mak-ro-*stem*-ma, long-stamened. The majority of fuchsias in cultivation are hybrids of this species.

PROCUMBENS, pro-*kum*-benz, prostrate.

RICCARTONI, rik-kar-*to*-ne, of the Riccarton Gardens, Edinburgh.

SPLENDENS, *splen*-denz, splendid.

TRIPHYLLA, trif-*il*-a, three-leaved—the leaves in threes round the stem.

**Fumaria,** few-*mar*-e-a; from L. *fumus,* smoke, some species having a smoky odour. Annual climber.

CAPREOLATA, kap-re-o-*la*-ta, tendrilled.

**Funkia,** *fungk*-e-a; after H. Funk, a German botanist. Herbaceous perennials.

FORTUNEI, for-*tu*-ne-i, after Robert Fortune, plant collector.

GRANDIFLORA, gran-dif-*lo*-ra, large flowered.

LANCIFOLIA, lan-sif-*o*-le-a, lance-leaved.

SIEBOLDIANA, se-bold-e-*a*-na, after Siebold.

SUBCORDATA, sub-kor-*da*-ta, somewhat heart-shaped—the leaves.

TARDIFLORA, tar-dif-*lo*-ra, late flowering.

UNDULATA, un-du-*la*-ta, waved—the leaves.

**Gagea,** *ga*-je-a; commemorating Sir Thomas Gage, a botanist. Hardy bulbs.

LUTEA, *loo*-te-a, yellow.

**Gaillardia,** gal-*lar*-de-a; after M. Gaillard, a French patron of botany. Hardy perennials; most garden kinds are hybrids.

AMBLYODON, am-*bly*-od-on, blunt toothed.

ARISTATA, ar-is-*ta*-ta, awned—the seeds.

PICTA, *pik*-ta, painted.

P. LORENZIANA, lor-enz-e-*an*-a, after

**Gaillardia** (*continued*)
Lorenz, German seedsman. A variety with tubular disk florets—"double" flowered in fact.

PULCHELLA, pul-*kel*-la, pretty (diminutive).

**Galanthus,** gal-*an*-thus; from Gr. *gala,* milk, and *anthos,* a flower, alluding to its whiteness—the Snowdrop. Bulbous.

BYZANTINUS, biz-an-*te*-nus. Byzantine.

ELWESII, el-*we*-zei, after Elwes, botanist and author.

NIVALIS, niv-*a*-lis, snowy. The Common Snowdrop.

**Galax,** *ga*-laks; from Gr. *gala,* milk, possibly in reference to the milk-white flowers. Woodland or peat plant.

APHYLLA, af-*il*-la, no leaves, that is, on the flower stalks.

**Galega,** ga-*le*-ga; from Gr. *gala,* milk, and *ago,* to lead, the plant once being esteemed as a fodder for cows and goats in milk, hence Goat's Rue. Hardy perennials.

OFFICINALIS, of-fis-in-*a*-lis, of the shop (herbal).

**Galium,** *ga*-le-um; from Gr. *gala,* milk, the leaves of G. verum once having been used for the curdling of milk. Rock plants.

OLYMPICUM, ol-*im*-pik-um, Mt. Olympus.

**Galtonia,** gawl-*to*-ne-a; commemorating Francis Galton, anthropologist. Hardy bulb.

CANDICANS, *kan*-dik-ans, white. The Spire Lily.

**Gamolepis,** gam-o-*lep*-is; after Gr. *gamos,* union, and *lepis,* a scale; refers to the scaly involucre. Half-hardy annual.

TAGETES, taj-*e*-tez, Tagetes (Marigold)-like.

**Gardenia,** gar-*de*-ne-a; commemorating Dr. Alexander Garden, a botanist of S. Carolina, U.S.A. Greenhouse shrubs.

FLORIDA, *flo*-rid-a (L. *floridus*), flowery.

**Garrya,** *gar*-re-a; named by Douglas in honour of Mr. Garry, of the Hudson Bay Company who gave the former much assistance in his plant-collecting expeditions in N.W. America. Shrubs.

ELLIPTICA, el-*lip*-tik-a, ellipse-shaped, the leaves.

**Gasteria**, gas-*te*-re-a; from Gr. *gaster*, belly, alluding to the swollen base of the flowers. Greenhouse succulents.

BREVIFOLIA, brev-*if*-ol-e-a, short-leaved.
DISTICHA, *dis*-tik-a, leaves in two rows.
LINGUA, *lin*-gwa, tongue-like—the leaves.
VERRUCOSA, ver-ru-*ko*-sa, warted.

**Gaultheria**, gawl-*the*-re-a; commemorating Dr. Gaulthier, a botanist and physician, of Quebec, in the eighteenth century. Shrubs.

NUMMULARIOIDES, num-mul-*ar*-e-oy-des, like Nummularia (Moneywort), the leaves and growths.
PROCUMBENS, pro-*cum*-benz, procumbent.
PYROLÆFOLIA, pir-ol-e-*fo*-le-a, leaves like Pyrola (Wintergreen).
SHALLON, *shal*-lon, old native American name.
TRICOPHYLLA, trik-of-*il*-la, hairy-leaved.

**Gaura**, *gaw*-ra; from Gr. *gauros*, superb. Hardy perennial.

LINDHEIMERI, lind-*hi*-mer-i, of Lindheimer, a botanist.

**Gaylussacia**, gay-loo-*sak*-e-a; named after a French chemist, J. L. Gay-Lussac. Peat shrubs.

DUMOSA, du-*mo*-sa, bushy.
FRONDOSA, fron-*do*-sa, leafy.
RESINOSA, rez-in-*o*-sa, resinous, the twigs and foliage.

**Gazania**, gaz-*a*-ne-a; from L. *gaza*, treasure, or riches, in allusion to the large and gaudy flowers, or commemorates Theodore of Gaza (d. 1478) who translated into Latin the botanical works of Theophrastus. Half-hardy perennials and annuals. The Treasure Flower.

LONGISCAPA, long-is-*ka*-pa, long-scaped —flower-stalk.
RIGENS, *ri*-gens, rigid.
SPLENDENS, *splen*-dens, splendid.

**Gelasine**, je-*las*-e-ne; from Gr. *gelasinos*, a smiling dimple, referring to the flowers. Bulbs.

AZUREA, az-*u*-re-a, azure blue.

**Genista**, jen-*is*-ta; ancient L. name. Flowering shrubs.

ÆTNENSIS, et-*nen*-sis, of Mount Etna.
CINEREA, sin-er-*e*-a, greyish—the foliage.
DALMATICA, dal-*mat*-ik-a, Dalmatian.
GLABRESCENS, glab-*res*-sens, slightly hairy.
HISPANICA, his-*pan*-ik-a, Spanish.
PILOSA, pil-*o*-sa, downy.

**Genista** (*continued*)

RADIATA, ra-de-*a*-ta, rayed—the form of the branches.
SAGITTALIS, sag-it-*ta*-lis, like an arrow— the winged twigs.
TINCTORIA, tink-*tor*-e-a, of dyers.
VIRGATA, ver-*ga*-ta, twiggy.

**Gentiana**, *jen*-she-*an*-a (or *jen*-te-*an*-a); called after Gentius, King of Illyria, who first used the plant in medicine. Rock and border perennials.

ACAULIS, a-*kaw*-lis, stemless.
ANDREWSII, *an*-drew-sei, after Andrews, a botanist of U.S.A.
ASCLEPIADEA, as-klep-e-*ad*-e-a, resembling the Asclepias.
BAVARICA, bav-*ar*-ik-a, of Bavaria.
FARRERI, *far*-rer-i, after Farrer, author and plant collector.
FREYNIANA, *freyn*-e-an-a, after Freyn.
KURROO, *ker*-roo, native name of habitat.
ORNATA, or-*na*-ta, ornate.
PNEUMONANTHE, new-mon-*an*-the, lit. lung-flower, the plant once being used as a remedy for lung disease.
PURDOMII, *pur*-dom-ei, after Purdom, the plant collector.
SEPTEMFIDA, sep-*tem*-fid-a, seven-cleft— the flowers.
SINO-ORNATA, *si*-no-or-*na*-ta, Chinese ornate gentian.
VERNA, *ver*-na, spring.

**Geonoma**, je-o-*no*-ma; from Gr. *geonomos*, skilled in agriculture, possibly because it can only be grown by a skilled cultivator. Warm-house palms.

GRACILIS, *gras*-il-is, slender.

**Geranium**, jer-*a*-ne-um; from Gr. *geranos*, a crane, the fruit of the plant resembling the head and beak of that bird, hence Cranesbill. Border perennials and rock plants. Greenhouse and bedding "geraniums" are strictly Pelargoniums, which see.

ALBANUM, al-*ba*-num, of Albania.
ARMENUM, ar-*me*-num, of Armenia.
ENDRESSII, en-*dres*-sei, after Endress.
GRANDIFLORUM, gran-dif-*lo*-rum, large-flowered.
IBERICUM, ib-*e*-rik-um, of Iberia (old name for Spain).
MACRORRHIZUM, mak-ror-*rhiz*-um, large rooted.
NEPALENSE, nep-al-*en*-se, of Nepal.
NODOSUM, no-*do*-sum, full of nodes, thick-jointed.
PHÆUM, *fe*-um, brownish or swarthy. The Dusky Cranesbill.

46

**Geranium** (*continued*)

PRATENSE, pra-*ten*-se, of meadows. The Meadow Craneshill.

PYLZOWIANUM, pil-zo-e-*an*-um, after Pylzow.

ROBERTIANUM, ro-ber- (or bair-) te-*an*-um, after Robert, a French abbot. The Herb Robert.

STRIATUM, stri-*a*-tum, streaked—the flowers.

TRAVERSI, tra-*ver*-se, after Travers.

WALLICHIANUM, wol-litsch-c-*a*-num, of Dr. Wallich, a director of Botanical Gardens, Calcutta.

**Gerardia**, jer-*ard*-e-a; named after John Gerard, who wrote his famous Herbal in the time of Queen Elizabeth. Annuals and perennials.

PURPUREA, pur-*pur*-e-a, purple coloured.

QUERCIFOLIA, kwer-kif-*ol*-e-a, Quercus (oak)-leaved.

**Gerbera**, *jer*-ber-a; after Gerber, a German naturalist. Tender perennials.

AURANTIACA, aw-ran-te-*a*-ka, golden-orange.

JAMESONII, jame-*so*-nei, after Jameson, a botanist. The Barberton Daisy.

**Gesnera**, *ges*-ner-a; after Conrad Gesner, a Swiss botanist. Tropical flowering and foliage tuberous plants.

CARDINALIS, kar-din-*a*-lis, red.

EXONIENSIS, eks-on-e-*en*-sis, of Exeter.

NÆGELIOIDES, næg-el-e-*oy*-dez, nægelia-like.

REFULGENS, ref-*ul*-jenz, shining.

**Geum**, *je*-um; possibly from Gr. *geuo*, to give an agreeable flavour (to taste), the roots of some species being aromatic. Herbaceous perennials.

BORISII, *bor*-is-ei, after Boris.

CHILOENSE, chil-o-*en*-se, of the Island of Chiloe.

HELDREICHII, hel-*drich*-ei, after Heldreich, botanist.

MONTANUM, mon-*ta*-num, of mountains.

REPTANS, *rep*-tanz, creeping.

RIVALE, re-*va*-le, pertaining to brooks.

**Gibbæum**, gib-*be*-um; from L. *gibba*, a hump, referring to the humped appearance of the larger of the two leaves forming the new growth. Greenhouse succulents.

ALBUM, *al*-bum, white—the flowers.

DISPAR, *dis*-par, unequal—the leaves.

GIBBOSUM, gib-*bo*-sum, humped or swollen.

PUBESCENS, pu-*bes*-cenz, downy.

**Gilia**, *gil*-e-a; after Felipe L. Gil, or Gilio, a Spanish botanist. Mainly half-hardy annuals.

CAPITATA, kap-it-*a*-ta, headed.

CORONOPIFOLIA, kor-on-op-if-*ol*-e-a, coronopus-leaved.

TRICOLOR, *trik*-ol-or, three-coloured.

**Gillenia**, gil-*le*-ne-a; named after Arnold Gille, German physician who wrote on horticulture in seventeenth century. Herbaceous perennial.

TRIFOLIATA, trif-ol-e-*a*-ta, leaves in threes.

**Ginkgo**, *gink*-go; Chinese name for the Maidenhair tree but regarded as incorrectly spelled.

BILOBA, bil-*o*-ba, two-lobed, the foliage.

**Gladiolus**, glad-e-*o*-lus; from Latin *gladiolus*, a little sword, the shape of the leaf. Some authorities give two pronunciations, glad-*i*-o-lus and glad-e-*o*-lus, the former being correct, but unlikely ever to be generally used. Bulbous plants, mainly hybrids in common cultivation.

BLANDUS, *bland*-us, pleasing or beautiful.

BRENCHLEYENSIS, brench-ley-*en*-sis, of Brenchley.

BYZANTINUS, biz-an-*te*-nus, Byzantine or Turkish.

CARDINALIS, kar-din-*a*-lis, red.

COLVILLEI, *col*-vil-*le*-i, after Colville.

GANDAVENSIS, gan-dav-*en*-sis, of Ghent.

PRIMULINUS, prim-ul-*e*-nus, primrose-coloured.

PSITTACINUS, sit-tak-*e*-nus, like a parrot—the colours of the flowers.

TRISTIS, *tris*-tis, sad—the mournful floral tint.

**Glaucium**, *glaw*-ke-um (or *glaw*-se-um); from Gr. *glaukos*, blue-grey, the colour of the foliage. Hardy biennials.

CORNICULATUM, kor-nik-ul-*a*-tum, horned. The Horned Poppy.

FLAVUM, *fla*-vum, yellow—the flowers.

**Gleditschia**, gled-*its*-she-a; named after Gottlieb Gleditsch, professor of botany in the eighteenth century. Trees.

CASPICA, *kas*-pik-a, Caspian.

TRIACANTHOS, tri-a-*kan*-thos, three-spined, *i.e.*, in threes.

**Gleichenia**, gle-*kee*-ne-a; after Baron Gleichen, a German botanist. Greenhouse ferns.

CIRCINATA, ser-sin-*a*-ta, coiled.

DICARPA, dik-*ar*-pa, two-fruited.

RUPESTRIS, roo-*pes*-tris, rock loving.

**Globularia**, glob-u-*lar*-e-a; from L. *globulus*, a small globe, in reference to the rounded flower heads. Sub-shrubs and perennials.

BELLIDIFOLIA, bel-lid-if-*o*-le-a, daisy-leaved.

CORDIFOLIA, kor-dif-*ol*-e-a, heart-shaped.

INCANESCENS, in-ka-*nes*-sens, becoming hoary.

TRICHOSANTHA, trik-o-*san*-tha, hairy-flowered—the appearance of the blossoms.

**Gloriosa**, glo-re-*o*-sa; from L. *gloriosus*, glorious, referring to the gorgeous flowers. Tropical climbing tuberous-rooted plants.

GRANDIFLORA, gran-dif-*lo*-ra, large flowered.

MAGNIFICA, mag-*nif*-e-ka, magnificent.

ROTHSCHILDIANA, roths-*child*-e-a-na, after Baron Rothschild.

SUPERBA, su-*per*-ba, superb.

VIRESCENS, ver-*es*-senz, green-flowered.

**Glottiphyllum**, glot-tif-*il*-lum; from Gr. *glotta*, the tongue, referring to the shape of the leaves. Greenhouse succulents.

APICULATUM, ap-ik-ul-*a*-tum, pointed—the leaves.

LINGUÆFORME, lin-gwe-*for*-me, tongue-shaped.

LONGUM, *long*-um, long—the leaves.

**Gloxinia**, gloks-*in*-e-a; named after P. B. Gloxin, a botanist of Colmar. Tuberous perennials. All well-known kinds are hybrids derived from—

SPECIOSA, spes-e-*o*-sa, showy.

**Glyceria**, glik-*er*-e-a (or gly-*ee*-re-a); from Gr. *glykys*, sweet—the foliage and roots. Aquatic grass.

AQUATICA, a-*kwak*-ik-a, growing in water.

CANADENSIS, kan-a-*den*-sis, of Canada.

**Glycyrrhiza**, glik-er-*rhe*-za (or gly-ser-*rhy*-za); from Gr. *glykys*, sweet, and *rhiza*, a root, referring to the sweet juice of the roots from which Spanish liquorice is made. Hardy herbaceous perennial.

GLABRA, *glab*-ra, smooth, destitute of hairs.

**Gnaphalium**, naf-*a*-le-um; from Gr. *gnaphalion*, woolly, in allusion to the very woolly foliage. Hardy herbaceous plants distributed under Anaphalis and Leontopodium.

**Gnidia**, *nid*-e-a; from Gnidus, a town in Crete. Greenhouse flowering shrubs.

DENUDATA, de-nu-*da*-ta, uncovered or not hairy.

PINIFOLIA, py-nif-*ol*-e-a, pine-leaved

**Godetia**, go-*de*-she-a (or god-*e*-te-a); after Charles H. Godet, Swiss botanist and entomologist. Hardy annuals. Often included under Œnothera.

GRANDIFLORA, gran-dif-*lo*-ra, large flowered.

SCHAMINII, *sham*-in-ei, after Schamin.

WHITNEYI, *whit*-ney-e, after Whitney.

**Goldfussia**, gold-*fus*-se-a; after Dr. Goldfuss, natural history professor in Bonn university. Greenhouse flowering plants.

ANISOPHYLLUS, an-is-of-*il*-us, unequal leaved or unsymmetrical.

ISOPHYLLA, is-of-*il*-a (or i-so-*fil*-a) equal leaved.

**Gomphrena**, gom-*free*-na; from an ancient name for an amaranth. Tender annuals.

GLOBOSA, glob-*o*-sa, globular. The Globe Amaranth.

**Goniophlebium**, gon-e-of-*leb*-e-um; from Gr. *gonia*, an angle, and *phleps*, a vein, alluding to the angled veins. Warmhouse ferns.

APPENDICULATUM, ap-pen-dik-ul-*a*-tum, drooping like an appendage.

SUBAURICULATUM, sub-aw-rik-ul-*a*-tum, somewhat eared—the fronds at their bases.

**Gordonia**, gor-*do*-ne-a; named after James Gordon, British nurseryman. Trees.

LASIANTHUS, las-i-*an*-thus, hairy-flowered.

PUBESCENS, pew-*bes*-senz, downy.

**Gossypium**, gos-*sip*-e-um; ancient name of the cotton plant. Warm-house plants.

BARBADENSE, bar-bad-*en*-se, from Barbados.

HERBACEUM, her-*ba*-se-um, herbaceous.

**Grammnanthes**, gram-*nan*-thes; from Gr. *gramma*, writing, and *anthos*, a flower, alluding to the letter-like marks on the petals. Half-hardy annuals.

GENTIANOIDES, jen-she-an-*oy*-des, gentian-like.

**Greenovia**, green-*o*-ve-a; named after George Ballas Greenough, a geologist. Greenhouse succulents.

AUREA, *aw*-re-a, golden—the flowers.

**Grevillea**, gre-*vil*-le-a; named after C. F. Greville, a patron of botany. Greenhouse and hardy shrubs.

SULPHUREA, sul-*fu*-re-a, sulphur-coloured —the flowers.

ROBUSTA, ro-*bus*-ta, strong or robust.

ROSMARINIFOLIA, ros-mar-*e*-nif-*ol*-e-a, rosemary (Rosmarinus)-leaved.

**Griffinia**, grif-*fin*-e-a; after W. Griffin, who brought these plants from Brazil. Tropical bulbous plants.

HYACINTHINA, hy-a-*sinth*-in-a, hyacinth-like.

**Griselinia**, gris-el-*in*-e-a; after Franc Griselini, a Venetian botanist. Shrub.

LITTORALIS, lit-tor-*a*-lis, of the sea-shore.

**Grusonia**, grus-*on*-e-a; after Hermann Gruson of Magdebourg, a cactus specialist. Greenhouse cacti.

CEREIFORMIS, *se*-re-if-*or*-mis, Cereus-like.

**Guaiacum**, gwa-e-*ak*-um; the native name. Tropical evergreen tree.

OFFICINALE, of-fis-in-*a*-le, found in shops, commercial.

**Gunnera**, *gun*-ner-a; named after J. E. Gunner, a botanist of Norway. Herbaceous perennials.

CHILENSIS, chil-*en*-sis, Chilian.

MAGELLANICA, maj-el-*an*-ik-a, of Magellan.

MANICATA, man-ik-*a*-ta, lit. sleeved, usually applied to pubescent matter which may be stripped off in shreds.

SCABRA, *sca*-bra, rough.

**Gymnocalycium**, gim-no-kal-*ik*-e-um; from Gr. *gymnos*, naked, and *kalya*, a bud, the flower buds having no covering. Greenhouse cacti.

DENUDATUM, de-nu-*da*-tum, denuded of covering.

GIBBOSUM, gib-*bo*-sum, humped.

LEEANUM, lee-*a*-num, after James Lee, the distinguished Hammersmith nurseryman who raised it from seed in 1840.

MULTIFLORUS, mul-tif-*lor*-us, many flowered.

**Gymnocladus**, gim-no-*kla*-dus; from Gr. *gymnos*, naked, and *klados*, a branch, presumably in allusion to the buds being almost invisible on the winter branches. Trees.

CANADENSIS, kan-a-*den*-sis, of Canada.

**Gymnogramma**, gim-no-*gram*-ma; from Gr. *gymnos*, naked, and *gramma*, writing, from the conspicuous characters formed by the branching sori. Greenhouse ferns. Gold and silver ferns.

**Gymnogramma** (*continued*)

ARGYROPHYLLUM, ar-ger-of-*il*-lum, silver-leaved.

CALOMELANOS, cal-om-*el*-an-os, beautiful black.

CHRYSOPHYLLA, kris-of-*il*-la, golden.

DEALBATA, de-al-*ba*-ta, whitewashed—the silvery farina on fronds.

JAPONICA, jap-*on*-ik-a, of Japan.

PERUVIANA, pe-ru-ve-*a*-na, of Peru.

TARTAREA, tar-*tar*-e-a, after Tartarus, "the deepest abyss of the infernal regions"; application to this South American fern is not very apparent.

WETTENHALLIANA, wet-ten-hal-le-*a*-na, after Wettenhall.

**Gynerium**, gyn-*eer*-e-um; from Gr. *gune*, ovary, and *erion*, wool, the stigmas being covered with wool.

ARGENTEUM, ar-*jen*-te-um, silvery. The Pampas Grass.

**Gypsophila**, gip-*sof*-il-la; from Gr. *gypsos*, chalk, and *phileo*, to love, in reference to the plant's preference for chalky soil. Border and rock perennials and annuals.

CERASTIOIDES, se-ras-te-*oy*-des, like Cerastium.

ELEGANS, *el*-e-ganz, elegant.

FRANKENIOIDES, fran-ken-e-*oy*-des, like Frankenia.

PANICULATA, pan-ik-ul-*a*-ta, panicled.

REPENS, *re*-penz, creeping.

**Habenaria**, hab-en-*ar*-e-a; from L. *habena*, a strap, referring to the strap-shaped spur. Terrestrial orchids.

BIFOLIA, bif-*ol*-e-a, two-leaved.

BLEPHARIGLOTTIS, blef-ar-e-*glot*-tis, eyelash-tongued—lip ciliated.

FIMBRIATA, fim-bre-*a*-ta, fringed, the flowers.

**Haberlea**, *hab*-er-le-a; named after Karl Konstantin Haberle, a botanist, of Pesth. Rock plants.

FERDINANDI-COBURGI, fer-de-*nan*-di-*ko*-burg-i, after scion of the Royal House of Austria.

RHODOPENSIS, ro-do-*pen*-sis, from Rhodope mountains.

**Habranthus**, hab-*ran*-thus; from Gr. *habros*, delicate, and *anthos*, a flower, of frail or delicate appearance. Half-hardy bulbs.

PRATENSIS, pra-*ten*-sis, of meadows.

**Habrothamnus**, hab-roth-*am*-nus; from Gr. *habros*, gay, and *thamnos*, a shrub, referring to the showy flowers. Greenhouse shrubs.

**Habrothamnus** (*continued*)
ELEGANS, *el*-e-ganz, elegant.
NEWELLI, new-*el*-le, after Newell.

**Hacquetia**, hak-*kwet*-e-a; named after B. Hacquet, a German botanist. Rock garden herb.
EPIPACTIS, ep-e-*pak*-tis, an old generic name.

**Hæmanthus**, ha-*man*-thus; from Gr. *haima*, blood, and *anthos*, a flower, alluding to red colour of flowers. Greenhouse bulbous plants.
ALBIFLOS, *al*-be-flos, white-flowered.
COCCINEUS, kok-*sin*-e-us, scarlet.
KATHARINÆ, kath-ar-*ee*-ne, after Catherine.
MULTIFLORUS, mul-tif-*lo*-rus, many-flowered.
NATALENSIS, nat-al-*en*-sis, of Natal.

**Hamatocactus**, ham-*a*-to-*kak*-tus; from L. *hamatus*, hooked, and *cactus*, refers to the hooked central spine. Greenhouse cactus.
SETISPINUS, se-tis-*pin*-us, spines bristle-like.

**Halesia**, hale-*e*-ze-a; commemorating Dr. Stephen Hales, a botanical author. Shrubs.
DIPTERA, *dip*-ter-a, two-winged—the seed pods.
TETRAPTERA, tet-*rap*-ter-a, four-winged—the seed pods.

**Halimodendron**, hal-e-mod-*en*-dron; from *halimus*, sea-coast, and *dendron*, a tree—a seaside shrub.
ARGENTEUM, ar-*jen*-te-um, silvery.

**Hamamelis**, ham-a-*me*-lis, from Gr. *hama*, together, and *mela*, fruit, flowers and fruit being borne at the same time. Shrubs.
ARBOREA, ar-*bor*-e-a, tree-like.
JAPONICA, jap-*on*-ik-a, Japanese.
MOLLIS, *mol*-lis, downy.
VIRGINIANA, vir-jin-e-*an*-a, of Virginia.

**Hardenbergia**, har-den-*ber*-je-a; after Countess of Hardenberg, sister to Baron Hugel, a plant collector and whose plants his sister cared for while on his travels. Greenhouse twining plants.
COMPTONIANA, kom-ton-e-*a*-na, after Compton.
MONOPHYLLA, mon-of-*il*-a, one-leafed—the leaflets reduced to one.

**Harpalium**, har-*pal*-e-um; from Gr. *harpaleos*, greedy, referring to vigour of growth. Herbaceous perennial.
RIGIDUM, *rig*-id-um, stiff or rigid stems.

**Hatiora**, hat-e-*or*-a; after Thomas Hariota—the name being an anagram—who was a botanist of the sixteenth century.
SALICORNOIDES, sal-ik-or-*noy*-dez, glasswort (Salicornia)-like.

**Haworthia**, ha-*worth*-e-a; after A. H. Haworth, an English botanist and writer on succulent plants. Greenhouse succulents.
ATROVIRENS, *a*-tro-*ver*-enz, dark green.
MARGARITIFERA, mar-gar-e-*tif*-er-a, pearl bearing—the white raised pimples on the leaves.
REINWARDTII, rin-*ward*-ei, after Reinwardt.

**Hebenstreitia**, he-ben-*stry*-te-a; after Professor J. E. Hebenstreit, of Leipzig. Half-hardy annual.
COMOSA, ko-*mo*-sa, hairy.

**Hedera**, *hed*-er-a; the ancient Latin name for Ivy.
ARBORESCENS, ar-bor-*es*-senz, tree-like.
CANARIENSIS, kan-ar-e-*en*-sis, of Canary Islands.
DIGITATA, dij-it-*a*-ta, fingered.
HELIX, *he*-liks, spiral, or twisted. The Common Ivy.
RÆGNERIANA, reg-ner-e-*an*-a, after Rægner.

**Hedychium**, he-*dik*-e-um; from Gr. *hedys*, sweet, and *chion*, snow, alluding to the snow-white fragrant flowers (of some species). Tropical herbaceous plants.
CORONARIUM, kor-on-*ar*-e-um, garlanded.
GARDNERIANUM, gard-ner-e-*a*-num, after Gardner.
MAXIMUM, *maks*-im-um, largest.

**Hedysarum**, hed-e-*sar*-um; name of obscure meaning, adopted from the Greek philosopher Theophrastus. Shrubby perennials.
CORONARIUM, kor-on-*ar*-e-um, crown or wreath-like—the flowers.
MULTIJUGUM, mul-te-*ju*-gum, many-paired leaves.

**Helenium**, hel-*e*-ne-um; possibly after Helen of Troy, a legend stating that these flowers sprang from her tears. Helen-flower. Herbaceous perennials.
AUTUMNALE, aw-tum-*na*-le, autumnal.

**Helenium** (*continued*)
   BIGELOVII, big-*lo*-vei, Prof. Bigelow, of Boston, U.S.A.
   BOLANDERI, bo-*land*-er-e, after Bolander.
   GRANDICEPHALUM, gran-de-*sef*-a-lum, large-headed.
   STRIATUM, stri-*a*-tum, striped or streaked.

**Helianthemum,** he-le-*an*-the-mum; from Gr. *helios*, the sun, and *anthemon*, a flower, the Sun Rose. Mainly dwarf shrubs. Also pronounced he-le-*anth*-em-um.
   FORMOSUM, for-*mo*-sum, beautiful.
   HALIMIFOLIUM, hal-im-if-*ol*-e-um, leaved like Halimus.
   OCYMOIDES, os-e-*moy*-dez, like Ocimum.
   VINEALE, vin-e-*a*-le, inhabiting vineyards.
   VULGARE, vul-*gar*-e, common. Many forms of this.

**Helianthus,** he-le-*an*-thus; from Gr. *helios*, the sun, and *anthos*, a flower, the Sunflower. Annuals and perennials.
   ANNUUS, *an*-u-us, annual.
   ARGOPHYLLUS, ar-*gof*-il-lus, silvery foliaged—leaves clothed in white down.
   CUCUMERIFOLIUS, ku-ku-mer-if-*ol*-e-us, Cucumis (cucumber)-like foliage.
   DECAPETALUS, dek-a-*pet*-al-us, ten-petalled.
   GIGANTEUS, ji-*gan*-te-us, gigantic.
   MULTIFLORUS, mul-tif-*lo*-rus, many-flowered.
   RIGIDUS, *rig*-id-us, rigid.
   SPARSIFOLIUS, spar-sif-*ol*-e-us, the leaves spread out, or far apart.
   TOMENTOSUS, to-men-*to*-sus, felted.

**Helichrysum,** he-le-*kri*-sum; from Gr. *helios*, the sun, and *chrusos*, gold, alluding to the flowers of some species. Border plants and shrubs.
   BELLIDIOIDES, bel-lid-e-*oy*-dez, daisy-like.
   BRACTEATUM, brak-te-*a*-tum, bracted—the flowers having many bracts.
   ROSMARINIFOLIUM, ros-mar-in-if-*ol*-e-um, Rosmarinum (rosemary)-leaved.
   TRINERVE, trin-*er*-ve, three-nerved—the leaves.

**Heliconia,** hel-ik-*o*-ne-a; from Gr. *Helicon*, a hill in Greece. Tropical herbaceous foliage plants.
   AUREO-STRIATA, *aw*-re-o-stre-*a*-ta, golden striped—the leaves.
   ILLUSTRIS, il-*lus*-tris, renowned.
   RUBRICAULIS, roob-rik-*aw*-lis, red stemmed.

**Heliocereus,** he-le-o-se-*re*-us; from Gr. *helios*, the sun, and *cereus*—flowers expand in sunlight. Greenhouse cacti.

**Heliocereus** (*continued*)
   SPECIOSUS, spes-e-*o*-sus, beautiful. An old plant known as Cereus speciosus and much used for hybridising with phyllocactus.

**Heliophila,** he-le-*of*-il-a; from Gr. *helios*, the sun, and *phileo*, to love. Half-hardy annuals.
   ARABOIDES, ar-ab-*oy*-des, arabis-like.
   LINEARIFOLIA, lin-e-ar-if-*ol*-e-a, narrow leaved.
   PILOSA, pil-*o*-sa, shaggy.

**Heliopsis,** he-le-*op*-sis; from Gr. *helios*, the sun, and *opsis*, a resemblance, the appearance of the flowers. Hardy perennials.
   LÆVIS, *le*-vis, smooth.
   SCABRA, *ska*-bra, rough.

**Heliotropium,** he-le-o-*tro*-pe-um; from Gr. *helios*, the sun, and *trope*, to turn, the flowers turning to the sun. Greenhouse sub-shrubs.
   PERUVIANUM, per-u-ve-*a*-num, from Peru.

**Helipterum,** hel-*ip*-ter-um; from Gr. *helios*, the sun, and *pteron*, a wing or feather, in reference to the plumed seed pappus; another derivation is L. *helix*, a screw, and *pteron*, a wing. Hardy annuals.
   MANGLESII, mang-*les*-ei, after Mangles.
   ROSEUM, *ro*-ze-um, rose-coloured.

**Helleborus,** hel-*le*-bor-us (popular pronunciation of Hellebore, *hel*-le-bore); also pronounced hel-*leb*-or-us; classical name of one of the species. Fibrous-rooted herbaceous perennials.
   CAUCASICUS, kaw-*kas*-ik-us, Caucasian.
   COLCHICUS, *kol*-chik-us, of Colchis.
   FŒTIDUS, *fet*-id-us, fetid.
   GUTTATUS, gut-*ta*-tus, spotted.
   LIVIDUS, *liv*-id-us, leaves bluish or lead-colour.
   NIGER, *ni*-jer, black—the root. Christmas Hellebore or Christmas Rose.
   VIRIDIS, *ver*-id-is, green—the flowers.

**Helonias,** hel-*o*-ne-as; from Gr. *helos*, a marsh, H. bullata being a bog plant.
   BULLATA, bul-*la*-ta, puckered—like a primrose leaf.

**Heloniopsis,** hel-o-ne-*op*-sis; from *Helonias* (*q.v.*), and Gr. *opsis*, a resemblance. Herbaceous plants.
   BREVISCAPA, brev-is-*ka*-pa, short-scaped (stalked).

51

**Helxine**, helks-in-*e* (pop., *helks*-een); from Gr. name for allied plant. Half-hardy carpeting or rock plant.

SOLEIROLII, so-le-er-*o*-lei, after Soleirol, a collector in China.

**Hemerocallis**, hem-er-o-*kal*-lis; from Gr. *hemeros*, a day, and *kallos*, beauty. Herbaceous plants. The Day Lily.

AURANTIACA, aw-ran-te-*a*-ka, golden-orange.

DUMORTIERI, du-mor-te-*a*-ri, after Dumortier, a Belgian botanist.

FLAVA, *fla*-va, yellow.

FULVA, *ful*-va, tawny.

GRAMINEA, gram-*in*-e-a, grassy-leaved.

MIDDENDORFFII, mid-den-*dorf*-ei, after Middendorff, a botanist.

**Hemionitis**, he-me-on-*i*-tis; from Gr. *hemionos*, a mule, the species being thought barren. Warm-house ferns.

CORDATA, kor-*da*-ta, heart-shaped.

PALMATA, pal-*ma*-ta, lobed like a hand.

**Hepatica**, he-*pat*-ik-a; from Gr. *hepar* liver, which the lobed leaves are supposed to resemble. Now joined with Anemone.

ANGULOSA, ang-ul-*o*-sa, leaves lobed or angled.

TRILOBA, tril-*o*-ba, leaves three-lobed.

**Heracleum**, her-*ak*-le-um; named after Hercules, who is said to have discovered the plant's medicinal virtues; or after *heracles*, a plant consecrated to Hercules. Biennials and perennials.

GIGANTEUM, ji-*gan*-te-um, gigantic.

MANTEGAZZIANUM, man-te-gats-e-*a*-num, after Mantegazzi.

SIBIRICUM, si-*bir*-ik-um, Siberian.

**Hereroa**, her-e-*ro*-a; after Herer in East Africa; the native habitat.

DOLABRIFORME, dol-*a*-brif-*or*-me, axe- or hatchet-shaped—the leaves.

GRANULATA, gran-ul-*a*-ta, granular or rough.

**Herniaria**, her-ne-*ar*-e-a; from L. *hernia*, rupture, for which the plant was a supposed remedy. Creeping or carpeting rock plant.

GLABRA, *gla*-bra, smooth.

**Herpestis**, her-*pes*-tis; from Gr. *herpestes*, creeping—the habit of the plant. Aquatic.

AMPLEXICAULIS, am-*pleks*-e-*cau*-lis, stem clasping—the leaves.

**Hesperis**, *hes*-per-is; from Gr. *hesperos*, evening, the Sweet Rockets being more fragrant at that time of day. Biennial and herbaceous plants.

MATRONALIS, mat-ro-*na*-lis, pertaining to a matron or dame. Old English name, Dame's Violet.

**Heteromeles**, he-ter-*om*-el-es; from Gr. *heteros*, variable, and *melon*, an apple, alluding to the variable character of the fruits. Shrubs.

ARBUTIFOLIA, ar-bew-tif-*ol*-e-a, arbutus-leaved.

**Heuchera**, *hoy*-ker-a; named after Professor J. H. Heucher, a German botanist. Rock and border perennials; garden kinds mainly hybrids.

AMERICANA, a-mer-ik-*a*-na, American.

BRIZOIDES, briz-*oy*-des, like Briza (Quaking Grass).

HISPIDA, *his*-pid-a, shaggy.

MICRANTHA, mi-*kran*-tha, small-flowered.

SANGUINEA, san-*gwin*-e-a, blood-red.

**Hexacentris**, hex-a-*ken*-tris; from Gr. *hex*, six, and *centron*, a spur, the stamens having six spurs between them. Tropical climbers.

MYSORENSIS, my-sor-*en*-sis, of Mysore.

**Hibbertia**, hib-*ber*-te-a; after G. Hibbert, a patron of botany. Greenhouse climbers.

DENTATA, den-*ta*-ta, toothed leaves.

VOLUBILIS, vol-*u*-bil-is, twining.

**Hibiscus**, hi-*bis*-kus; name of very ancient origin used by Virgil for a mallow-like plant. Shrubs, perennials and annuals.

AFRICANUS, af-rik-*a*-nus, African.

MANIHOT, *man*-e-hot, Australian name.

ROSA-SINENSIS, *ro*-za-sin-*en*-sis, Chinese-rose.

SYRIACUS, syr-e-*ak*-us, Syrian, a misleading name, since this shrub is native only of China and India.

TRIONUM, tre-*o*-num, three-coloured. The Flower of an Hour Hibiscus or Bladder Ketmia.

**Hidalgoa**, hid-*al*-go-a; after Hidalgo, Mexican gentleman. Half-hardy climber.

WERCKLEI, *werk*-le-i, after Carlo Werckle.

**Hieracium**, hi-er-*ak*-e-um; an ancient name from Gr. *hierax*, a hawk. Pliny, the Roman naturalist, believed hawks ate the plant to strengthen their eyesight.

**Hieracium** (*continued*)
Also pronounced hi-er-*a*-se-um. The Hawkweed.

AURANTIACUM, aw-ran-te-*a*-kum, golden orange.
BORNMULLERI, born-*mool*-ler-i, after Bornmuller.
VILLOSUM, vil-*lo*-sum, covered with long, loose hairs.

**Hierochloe**, hi-er-o-*klo*-e; from Gr. *hieros*, sacred, and *chloa*, grass, alluding to this grass being strewn on church floors, hence the English name of Holy Grass.

BOREALIS, bor-e-*a*-lis, northern.

**Hippeastrum**, hip-pe-*as*-trum; from Gr. *hippeus*, a knight, and *astron*, a star, from some fancied resemblance to knights and stars (Knight's star) in H. equestre. Greenhouse bulbs, mainly of hybrid origin.

EQUESTRE, e-*kwes*-tree, equestrian.
PRATENSE, pra-*ten*-se, of meadows.
RETICULATUM, re-tik-ul-*a*-tum, colours forming a network.

**Hippocrepis**, hip-po-*kre*-pis; from Gr. *hippos*, a horse, and *crepis*, a shoe, in reference to the shape of the seed pod, which resembles a horseshoe. Rock plants.

COMOSA, kom-*o*-sa, hairy, in tufts.

**Hippophæ**, hip-*po*-fa-e; Gr. name for a spring plant. Shrub.

RHAMNOIDES, ram-*noy*-dez, like Rhamnus (buckthorn). The Sea Buckthorn.

**Hippuris**, hip-*pur*-is; from Gr. *hippos*, a horse, and *oura*, a tail—the stems, crowded with hair-like leaves, resemble a horse's tail. Submerged aquatic.

VULGARIS, vul-*gar*-is, common. The Mare's-tail.

**Hoffmannia**, hof-*man*-e-a; named after Georg Franz Hoffmann, professor of Göttingen. Herbs or woody plants with very showy foliage.

DISCOLOR, *dis*-ko-lor, different colours.
REGALIS, re-*ga*-lis, royal.

**Holbœllia**, hol-*bel*-le-a; after F. L. Holböll of Copenhagen Botanic Gardens. Greenhouse climber.

LATIFOLIA, lat-if-*ol*-e-a, broad-leaved.

**Holcus**, *hol*-kus; from Gr. *holkus*, the name of a grass. Hardy variegated grass.

LANATUS ALBO-VARIEGATUS, la-*na*-tus al-bo-var-e-eg-a-tus, white variegated soft-leaved.

**Homeria**, ho-*me*-re-a; said to be from *homereo*, alluding to the meeting or joining of the filament. Greenhouse bulbous plants.

COLLINA, kol-*le*-na, found on hills.

**Hoodia**, *hood*-e-a; a commemorative name. Greehouse succulents.

BAINII, *bane*-ei, after Bain.
GORDONI, *gor*-don-e, after Gordon.

**Hordeum**, *hor*-de-um; Latin name for Barley. Annual grasses.

JUBATUM, joo-*ba*-tum, crested or maned.

**Hosackia**, hos-*ak*-e-a; commemorating Dr. Hosack, an American botanist. Rock and border plants.

OBLONGIFOLIA, ob-long-if-*o*-le-a, oblong-leaved.
PURSHIANA, pursh-e-*an*-a, after Pursh, an American botanist.

**Hosta**, *hos*-ta; after Nicolous Thomas and Joseph Host, Austrian botanists. Hardy foliage and flowering plants.

This name is a Continental synonym of Funkia, the name in common use in English gardens for the Plantain Lily. The species will accordingly be found under Funkia.

**Hottonia**, hot-*to*-ne-a; after P. Hotton, a Dutch professor of botany. Aquatics.

PALUSTRIS, pal-*us*-tris, of marshes.

**Houstonia**, hows-*to*-ne-a; commemorating Dr. W. Houston, an English botanist. Rock plants.

SERPYLLIFOLIA, ser-pil-if-*ol*-e-a, leaves like wild thyme, Thymus serpyllum.

**Houttuynia**, howt-too-*in*-e-a; after Dr. Martin Houttuyn, of Amsterdam. Marsh plants.

CORDATA, kor-*da*-ta, heart-shaped, the leaves.

**Hovea**, *ho*-ve-a; after A. P. Hove, a Polish botanist. Greenhouse flowering shrub.

CELSII, *kel*-sei, after Cels.

**Howea**, *how*-e-a; after Lord Howe's Island, where these palms are found. Greenhouse and room palms.

BELMOREANA, bel-mo-re-*a*-na, after Belmore.
FORSTERIANA, fors-ter-e-*a*-na, after Forster.

**Hoya**, *hoy*-a; after Thomas Hoy, gardener at Sion House, seat of Duke of Northumberland. The Wax Flower.

BELLA, *bel*-la, pretty.

**Hoya** (*continued*)

CARNOSA, kar-*no*-sa, fleshy or flesh coloured—the flowers.

PAXTONII, *pax*-ton-ei, after Sir Joseph Paxton.

**Humea,** *hu*-me-a; after Lady Hume. Greenhouse biennial.

ELEGANS, *el*-e-ganz, elegant.

**Humulus,** *hu*-mul-us; origin uncertain, possibly from Latinised form of old German *humela*, hops, from L. *humus*, the ground, *i.e.*, lowly or trailing if unsupported. Climbing or twining plants.

JAPONICUS VARIEGATUS, jap-*on*-ik-us var-e-eg-*a*-tus, variegated hop of Japan.

LUPULUS, *loo*-pul-us, the old herbalists' shop name for the Hop, meaning a wolf; also applied to hook-like teeth with which the stems are armed and by which it climbs or clings to a support.

**Hunnemannia,** hun-ne-*man*-ne-a; after J. Hunnemann, a botanical traveller. Half-hardy perennial.

FUMARIÆFOLIA, fu-mar-e-e-*fo*-le-a, leaves like those of Fumaria (fumitory).

**Hutchinsia,** hut-*chin*-se-a; after Miss Hutchins, an Irish botanist. Rock plants.

ALPINA, al-*pine*-a, alpine.

**Hyacinthus,** hi-a-*sin*-thus or hi-ak-*in*-thus; named after Hyakinthos, the beautiful Spartan (Greek mythology), who was accidentally killed by Apollo, the legend stating that hyacinths sprang up where his blood was shed. Florists' hyacinths are forms or hybrids of H. orientalis and others.

AMETHYSTINUS, am-e-this-*te*-nus, amethyst.

AZUREUS, az-*u*-e-us, azure-blue.

ORIENTALIS, or-e-en-*ta*-lis, eastern.

ROMANUS, ro-*ma*-nus, Roman.

**Hydrangea,** hy-*dran*-je-a; from Gr. *hydor*, water, and *aggeion*, a vessel, or vase, in reference to the shape of the seed capsule. Shrubs.

ARBORESCENS, ar-bor-*es*-sens, tree-like.

HORTENSIS, hor-*ten*-sis, lit. of gardens, but it is stated that this name, applied to the common Hydrangea, commemorates Madame Hortense Lepante, wife of a celebrated clockmaker, of Paris. Hortensia was once the generic name of the species.

PANICULATA, pan-ik-ul-*a*-ta, panicled.

PETIOLARIS, pet-e-o-*lar*-is, long-petioled (leaf-stalk).

VESTITA, ves-*te*-ta, clothed with hairs.

**Hydrocharis,** hi-*drok*-ar-is; from Gr. *hydor*, water, and *charis*, graceful, in allusion to the beauty of the floating flowers. Aquatic.

MORSUS-RANÆ, mor-sus-*ra*-ne; from L. *morsus*, a bite, and *rana*, a frog. The Frog-bit.

**Hydrocotyle,** hi-dro-*kot*-il-e; from Gr. *hydor*, water, and *kotyle*, a cup, in reference to the cup-like hollow at the centre of the leaf. Bog plants.

VULGARIS, vul-*gar*-is, common.

**Hydropeltis,** hi-dro-*pel*-tis; from Gr. *hydor*, water, and *pelte*, a buckler or shield—habitat of plant and shape of leaves, Aquatic. The Water Shield.

PURPUREA, pur-*pur*-e-a, purple flowers.

**Hylocereus,** hi-lo-*se*-re-us; after Gr. *hyle*, a wood, and *cereus*; the habitat—epiphytic on trees. Greenhouse climbing cacti.

TRIANGULARIS, tri-ang-ul-*ar*-is, three-angled—the stem.

**Hymenanthera,** hi-men-*an*-ther-a; from Gr. *hymen*, a membrane, and *anthera*, an anther, in reference to the construction of the pollen bags. Shrubs.

CRASSIFOLIA, kras-sif-*ol*-e-a, thick-leaved.

**Hymenocallis,** hi-men-o-*kal*-is; from Gr. *hymen*, a membrane, and *kalos*, beautiful, a reference to the membraneous cup forming the flower centre.

CALATHINA, cal-ath-*ee*-na, basket flowered.

MACROSTEPHANA, mak-ros-*tef*-an-a, large crowned.

OVATA, o-*va*-ta, ovate.

SPECIOSA, spes-e-o-sa, showy.

**Hymenophyllum,** hi-men-of-*il*-lum; from Gr. *hymen*, a membrane, and *phyllon*, a leaf, alluding to the delicate membraneous fronds. Greenhouse filmy ferns.

DEMISSUM, dem-*is*-sum, pressed down, dwarf.

FLABELLATUM, flab-el-*la*-tum, fan-shaped.

PULCHERRIMUM, pul-*ker*-re-mum, very beautiful.

TUNBRIDGENSE, tun-brij-*en*-se, of Tunbridge Wells.

**Hypericum,** hi-*per*-ik-um (in classical Greek, hi-per-*e*-kum); Gr. name of obscure meaning; some authorities derive word from Gr. *hyper*, over, and *ereike*, a heath, possibly in reference to the natural

**Hypericum** (*continued*)
habitat of some species. Rock and border plants and shrubs.

ANDROSÆMUM, an-dros-*e*-mum, old generic name from Gr. *aner*, a man, and *aima*, blood, the berry being red. The Tutsan.

AUREUM, *aw*-re-um, golden.

BALEARICUM, bal-e-*ar*-ik-um, of the Balearic Isles.

CALYCINUM, kal-is-*e*-num, L. *calyx*, a cup, probably in allusion to the large cup-shaped calyx.

CORIS, *kor*-is, the leaves resembling Coris.

EMPETRIFOLIUM, em-pet-rif-*o*-le-um, leaved like Empetrum.

FRAGILE, *fraj*-il-e, fragile.

HIRCINUM, her-*se*-num, smelling of goats.

HOOKERIANUM, hook-er-e-*a*-num, after Hooker.

HUMIFUSUM, hu-me-*few*-sum, on the ground, prostrate.

HYSSOPIFOLIUM, his-sup-e-*fo*-le-um, leaved like Hyssopus.

KALMIANUM, kal-me-*an*-um, after Peter Kalm, Swedish naturalist and traveller, who discovered it.

LYSIMACHIOIDES, lis-im-ak-e-*oy*-des, like Lysimachia, presumably one of the yellow-flowered Loosetrifes.

MOSERIANUM, mo-ser-e-*a*-num, after Moser, a French nurseryman, who raised this hybrid.

PATULUM, *pat*-u-lum, spreading.

POLYPHYLLUM, pol-if-*il*-lum, many-leaved.

REPENS, *re*-penz, creeping and rooting.

REPTANS, *rep*-tans, creeping.

**Hypolepis,** hi-po-*lep*-is; from Gr. *hypo*, under, and *lepis*, a scale—the appearance of the fructification. Greenhouse ferns.

DISTANS, *dis*-tanz, distant, the frond divisions.

MILLEFOLIUM, mil-lef-*ol*-e-um, milfoil-leaved.

REPENS, *re*-penz, creeping and rooting.

**Hyssopus,** his-*so*-pus; a very ancient name used by the Greeks, probably of Hebrew origin. A pot herb. The Hyssop.

OFFICINALIS, of-fis-in-*a*-lis, of the shop (herbal).

**Iberis,** *i*-ber-is; from Iberia, the ancient name for Spain, where many species are common. Rock and border plants. Annuals and perennials.

CORIFOLIA, kor-if-*ol*-e-a, Coris-leaved.

CORONARIA, kor-on-*ar*-e-a, crown flowering. The Rocket Candytuft.

**Iberis** (*continued*)

CORREÆFOLIA, kor-e-e-*fo*-le-a, leaved like Correa.

GIBRALTARICA, jib-ral-*tar*-ik-a, of Gibraltar.

PRUITII, *pru*-it-ei, after Pruit, a botanist.

SAXATILIS, saks-*a*-til-is, inhabiting rocks.

SEMPERFLORENS, sem-per-*flor*-ens, always flowering.

SEMPERVIRENS, sem-per-*veer*-enz, always green.

TENOREANA, ten-or-e-*a*-na, after Professor Tenore, an Italian botanist.

UMBELLATUS, um-bel-*la*-tus, flowers in an umbel. Annual Candytuft.

**Idesia,** i-*de*-ze-a; after E. Y. Ides, a Dutchman, who travelled in China in the seventeenth century. Shrubs.

POLYCARPA, pol-ik-*ar*-pa, many-fruited.

**Ilex,** *i*-leks; from the old Latin name *ilex*, an evergreen oak (Holm Oak), to which the Holly was supposed to bear some resemblance. Many of the hollies in cultivation are forms of I. Aquifolium, while others are forms of I. platyphylla.

AQUIFOLIUM, ak-we-*fo*-le-um, old name, meaning pointed leaves.

CRENATA, kre-*na*-ta, leaves crenulate, *i.e.*, rounded teeth.

PERNYI, *per*-ne-i, after Abbé Perny, who discovered it.

PLATYPHYLLA, plat-e-*fil*-la, broad-leaved.

**Illicium,** il-*lis*-e-um; from L. *illicio*, to attract or allure, referring to the aromatic perfume. The Aniseed Tree.

ANISATUM, an-is-*a*-tum, anise-scented.

FLORIDANUM, flor-id-*a*-num, of Florida.

**Imantophyllum,** im-ant-of-*il*-lum; from Gr. *imas*, a leather thong, and *phyllon*, a leaf, alluding to the leaves which are strap-like in form and texture. An excellent name dropped in favour of the merely complimentary one of Clivia (which see). Both names were published on the same date, namely, October 1, 1828. Also spelled Imatophyllum.

**Impatiens,** im-*pa*-she-ens (or im-*pat*-e-ens); from L. *impatiens*, impatient or hasty, in allusion to the manner in which the pods of some species explode and scatter their seed when touched. Annuals and biennials.

BALSAMINA, bawl-sa-*me*-na, Balsam, old name for the group.

GLANDULIFERA, glan-dul-*if*-er-a, gland-bearing—the leaves.

## Impatiens (*continued*)

LONGICORNU, long-e-*kor*-nu, long-horned—the flowers.

NOLI-ME-TANGERE, no-li-me-*tan*-ger-e, pop. "touch me not," in reference to the expulsive action of the seed pods.

ROYLEI, *roy*-le-i, after Royle, a professor of botany.

SULTANI, sul-*tan*-i, after the Sultan of Zanzibar.

**Incarvillea**, in-kar-*vil*-le-a; commemorating P. Incarville, French Jesuit missionary to China. Herbaceous perennials.

BREVIPES, *brev*-e-pes, short-stalked.

DELAVAYI, del-a-*va*-i, after Delavay.

OLGÆ, *ol*-ge, commemorating a Princess Olga.

**Indigofera**, in-dig-*of*-er-a; from *indigo*, the blue dye (L. *indicus*, Indian, whence it comes), and *fero*, to produce. Shrubs.

GERARDIANA, jer-ar-de-*a*-na, after Gerard.

**Inula**, *in*-u-la; believed to be a corruption of *helenium*, I. helenium (Elecampane) being the *Inula campana* of medieval Latin. Herbaceous perennials.

ENSIFOLIA, en-sif-*ol*-e-a, sword-leaved.

GLANDULOSA, glan-dul-*o*-sa, glanded.

HELENIUM, hel-*e*-ne-um, old generic name, meaning Helen-flower.

MACROCEPHALA, mak-ro-*sef*-a-la, large-headed.

MONTANA, mon-*ta*-na, of mountains.

OCULUS-CRISTI, *ok*-u-lus-*kris*-te, Christ's eye, presumably the appearance of the blossom.

ROYLEANA, roy-le-*a*-na, after Royle, a botanist.

SALICINA, sal-is-*e*-na, Salix (willow)-like —the leaves.

**Ionopsidium**, i-on-op-*sid*-e-um; from Gr. *ion*, a violet, *opsis*, appearance, and *idion*, diminutive, *i.e.*, plant resembling a little violet. Dwarf plants.

ACAULE, ak-*aw*-le, stalkless—the plants.

**Ipomea**, ip-o-*me*-a; from Gr. *ips*, bindweed, and *homoios*, like, referring to the twining habit. Mostly annuals.

BATATAS, bat-*a*-tas, native name for Sweet Potato.

COCCINEA, kok-*sin*-e-a, scarlet.

HEDERACEA, hed-er-*a*-se-a, ivy-leaved.

PURPUREA, pur-*pur*-e-a, purple.

QUAMOCLIT, kwa-*mok*-lit, Quamoclit, the native name.

RUBRO-CŒRULEA, *roob*-ro-ser-*u*-le-a, red and blue.

VERSICOLOR, ver-*sik*-o-lor, changeable colour.

**Iresine**, i-res-*in*-e; from Gr. *eiros*, wool, alluding to the flowers and seeds. Coloured-leaved tender bedding plants.

HERBSTII, *herb*-stei, after Herbst; aureo-reticulata, a gold net-veined variety.

LINDENII, *lin*-den-ei, after Linden.

**Iris**, *i*-ris; from Gr. *iris*, a rainbow, presumably in reference to the many colours of the flowers. Bulbous, rhizomatous, and herbaceous perennials.

BULLEYANA, bul-le-*a*-na, after A. K. Bulley.

CHAMÆIRIS, kam-e-*i*-ris, on the ground, *i.e.*, dwarf.

CHRYSOGRAPHES, kris-o-*graf*-es, veined with gold.

CRISTATA, kris-*ta*-ta, crested.

DELAVAYI, del-a-*va*-i, after Delavay.

DOUGLASIANA, dug-las-e-*a*-na, after Douglas, botanist and explorer.

FILIFOLIA, fil-if-*ol*-e-a, thread-leaved, foliage narrow.

FLORENTINA, flor-en-*te*-na, of Florence.

FŒTIDISSIMA, fet-id-*is*-sim-a, most fetid —the leaf odour.

FOLIOSA, fo-le-*o*-sa, well-foliaged or leafy.

FORRESTII, for-*res*-tei, after Forrest.

FULVA, *ful*-va, tawny.

GERMANICA, jer-*man*-ik-a, of Germany. Flag Iris.

GRACILIPES, gras-*il*-e-pez, slender-stalked.

GRAMINEA, gram-*in*-e-a, grassy, the foliage.

HEXAGONA, heks-ag-*o*-na, the ovary six-angled.

HISTRIO, *his*-tre-o, L. an actor, the flower being very gaily adorned.

HOOGIANA, hoo-ge-*a*-na, after Hoog, a Dutch nurseryman.

JAPONICA, jap-*on*-ik-a, of Japan.

JUNCEA, *jun*-se-a, a rush, plant having rush-like leaves.

KÆMPFERI, *kem*-fer-i, after Kæmpfer, a German physician and traveller of the fifteenth century.

LACUSTRIS, la-*kus*-tris, a lake, found by the Great Lakes.

LÆVIGATA, lev-e-*ga*-ta, smooth.

MISSOURIENSIS, mis-soor-e-*en*-sis, found at source of River Missouri, Rocky Mountains.

MONNIERI, mon-ne-*air*-i, after Monnier, a French botanist.

OCHROLEUCA, ok-ro-*lew*-ka, yellow-white.

PALLIDA, *pal*-lid-a, pale, the flowers paler in colour than those of the commoner I. germanica.

PSEUDACORUS, sued-ak-*or*-us, false acorus so named to distinguish it from Acorus Calamus, true Acorus.

**Iris** (*continued*)

PUMILA, *pew*-mil-a, dwarf or diminutive.
RETICULATA, ret-ik-ul-*a*-ta, netted—the bulb.
RUTHENICA, roo-*then*-ik-a, Russian.
SIBIRICA, si-*bir*-ik-a, of Siberia.
SPURIA, *spu*-re-a, spurious (lit. a bastard), reason for so naming unknown.
STYLOSA, stil-*o*-sa, long-styled.
TECTORUM, tek-*tor*-um, on roofs, this iris being grown on thatched roofs in Japan.
TENAX, *te*-naks, tough—the fibres of the leaves.
TINGITANA, ting-e-*ta*-na, of Tangiers.
UNGUICULARIS, un-gwik-ul-*ar*-is, narrow-clawed (the lower end of the petal).
VERSICOLOR, ver-*sik*-o-lor, variable colour.
XIPHIOIDES, *zif*-e-oy-dez, xiphium-like. The Spanish Iris.
XIPHIUM, *zif*-e-um, old Greek name for Gladiolus segetum, reason unknown for its use here. The English Iris.

**Isatis,** i-*sa*-tis; classical name for a healing herb. Biennials.

GLAUCA, *glaw*-ka, blue-green—the foliage.
TINCTORIA, ting-*tor*-e-a, of dyers.

**Ismene,** is-*me*-ne; after *Ismene*, daughter of Œdipus and Jocasta in Greek legend. The genus is now joined to Hymenocallis.

**Isoetes,** is-o-*ee*-tez; from Gr. *isos*, equal, and *etos*, the year—the plant does not alter with the seasons. Submerged aquatic cryptogam. The Quillwort.

LACUSTRIS, la-*kus*-tris, found in lakes.

**Isolepis,** is-o-*lep*-is; from Gr. *isos*, equal, and *lepis*, a scale—the scales of the perianth are equal. Also pronounced i-so-*lep*-is. Greenhouse dwarf sedge.

GRACILIS, *gras*-il-is, slender.

**Isoloma,** is-o-*lo*-ma; from Gr. *isos*, equal, and *loma*, a border—the lobes of the corolla are equal. Warm-house flowering herbaceous plants.

DIGITALIFLORUM, dij-it-al-if-*lo*-rum, foxglove-like as to flowers.
HONDENSE, *hon*-den-se, of Honda, New Granada.

**Isopyrum,** is-o-*pi*-rum; from Gr. for like and wheat, as the seeds resemble those of wheat. Herbaceous perennials.

FUMARIOIDES, fu-mar-e-*oy*-des, Fumaria (fumitory)-like.
THALICTRIOIDES, thal-ik-tre-*oy*-dez, Thalictrum-like.

**Itea,** *i*-te-a; the old Greek name for a willow, from shape of leaves of one of the species or the adaptability of some species for damp soils. Shrubs.

ILICIFOLIA, il-is-if-*ol*-e-a, holly-leaved.

**Ixia,** *iks*-e-a; from the Gr. name *ixia*, birdlime, in reference to the sticky nature of the juice. Half-hardy bulbs. Mainly hybrids.

CRATEROIDES, kra-ter-*oy*-dez, crater-like.
VIRIDIFLORA, ver-id-if-*lo*-ra, green flowered.

**Ixiolirion,** iks-e-o-*lir*-e-on; from Gr. *ixia* (bird lime), and *leirion*, a lily, lit. the ixia-like lily. Hardy bulbs.

MONTANUM, mon-*ta*-num, of mountains.
TATARICUM, tah-*tar*-ik-um, from Tartary.

**Ixora,** iks-*or*-a; after Iswara, a Malabar deity to whom the flowers were offered. Tropical flowering shrubs.

COCCINEA, kok-*sin*-e-a, scarlet.
DUFFII, *duf*-fei, after Duff.

**Jacaranda,** jak-ar-*an*-da; the Brazilian name. Tropical trees grown as ornamental foliage pot plants.

CÆRULEA, se-*ru*-le-a, sky-blue.
MIMOSÆFOLIA, mim-*o*-se-*fol*-e-a, mimosa-leaved.

**Jacobinia,** jak-o-*bin*-e-a; derivation unknown. Greenhouse flowering shrubs.

CHRYSOSTEPHANA, kris-*os*-tef-*a*-na, golden crowned.
COCCINEA, kok-*sin*-e-a; scarlet.
GHIESBREGHTIANA, ghiez-brek-te-*a*-na, after Ghiesbreght.

**Jamesia,** *jame*-se-a; called after an American botanist, Dr. Edwin James, who first discovered it. Shrub.

AMERICANA, a-mer-ik-*a*-na, of America.

**Jasione,** jas-e-*o*-ne; ancient name, application uncertain. Rock plants.

HUMILIS, *hu*-mil-is, of low growth.
PERENNIS, per-*en*-nis, perennial.

**Jasminum,** jas-*min*-um; said to be derived from *ysmyn*, the Arabic name for Jasmine. Shrubby flowering climbers.

BEESIANUM, bee-ze-*a*-num, after Messrs. Bees, the nurserymen who introduced it.
FRUTICANS, *frut*-ik-anz, shrubby.
NUDIFLORUM, nu-dif-*lo*-rum, naked-flowered, the shrub blooming when the branches are leafless. The Winter Jasmine.
OFFICINALE, of-fis-in-*a*-le, of the shop (herbal). The Jessamine or Jasmine.

**Jeffersonia,** jef-fer-*so*-ne-a; commemorating Thomas Jefferson, once President of U.S.A. Woodland herb.
DIPHYLLA, dif-*il*-la, two-leaved, *i.e.*, in pairs.

**Juglans,** *jug*-lans; old Latin name for the Walnut tree, probably from *Jovis glans,* the nut of Jupiter in mythology. Trees.
REGIA, *re*-je-a, royal, application uncertain. The Walnut.

**Juncus,** *jun*-kus; from L. *jungo,* to bind or tie, the stems being used as cord. Bog plants.
EFFUSUS, ef-*few*-sus, spread-out—the leaves.
ZEBRINUS, ze-*bry*-nus, zebra-striped—the leaves.

**Juniperus,** ju-*nip*-er-us; old Latin name for the Juniper tree. Evergreen trees.
CHINENSIS, tshi-*nen*-sis, of China.
COMMUNIS, kom-*mu*-nis, common, *i.e.*, in groups or communities. The Juniper.
DRUPACEA, droo-*pa*-se-a, alluding to the drupe-like fruit.
EXCELSA, eks-*sel*-sa, tall.
PACHYPHLÆA, pak-e-*fle*-a, thick-barked.
PROCUMBENS, pro-*kum*-benz, procumbent.
SABINA, sa-*bi*-na, old Latin name for the Savin.
VIRGINIANA, vir-jin-e-*a*-na, of Virginia.

**Jussieua,** jus-*su*-a (also Jussiæa, jus-*se*-e-a); after Bernard de Jussieu, of Paris. Bog aquatics.
LONGIFOLIA, long-if-*ol*-e-a, long-leaved.
REPENS, *re*-penz, creeping and rooting.
SPRENGERI, *spreng*-er-e, after Sprenger.

**Justicia,** jus-*tis*-e-a; after James Justice, F.R.S., a noted Scottish gardener. Greenhouse flowering shrubs.
CALYCOTRICHA, kal-ik-*o*-trik-a, beautiful haired.
CARNEA, *kar*-ne-a, flesh-coloured.
FLAVICOMA, flav-e-*ko*-ma; yellow haired.

**Kæmpferia,** keem-*feer*-e-a; after Kæmpfer, a traveller in Japan. Tropical foliage plants.
GILBERTII, *gil*-bert-ei, after Gilbert.
ROTUNDA, ro-*tund*-a, round—as to roots.

**Kalanchoe,** kal-an-*ko*-e; the native name for a Chinese species. Greenhouse succulents.
CARNEA, *kar*-ne-a, flesh-coloured.
DYERI, *dy*-er-i, after Dyer.
FLAMMEA, *flam*-me-a, flame-coloured or fiery red.

**Kalanchoe** (*continued*)
KEWENSIS, kew-*en*-sis, of Kew Gardens, where raised.
MARMORATA, mar-mor-*a*-ta, marbled-leaved.

**Kalmia,** *kal*-me-a; named after Peter Kalm, one of the pupils of Linnæus. Flowering shrubs.
ANGUSTIFOLIA, an-gus-tif-*o*-le-a, narrow-leaved.
GLAUCA, *glaw*-ka, bluish-white—leaf underparts.
LATIFOLIA, lat-if-*o*-le-a, broad-leaved.

**Kalosanthes,** kal-os-*an*-theez; from Gr. *kalos,* beautiful, and *anthos,* a flower, in reference to floral beauty. Greenhouse succulent flowering plants.
COCCINEA, kok-*sin*-e-a, scarlet.

**Kaulfussia,** kaul-*fus*-se-a; commemorating a Dr. G. F. Kaulfuss, a German botanist. Hardy annual.
AMELLOIDES, am-el-*loy*-dez, like Aster Amellus.

**Kennedya,** ken-*ned*-e-a; after Mr. Kennedy, a nurseryman. Greenhouse flowering climbers.
COCCINEA, kok-*sin*-e-a, scarlet.
MARRYATTÆ, mar-re-*at*-te, after Mrs. Marryatt.
PROSTRATA, pros-*tra*-ta, prostrate, lying flat.

**Kentia,** *kent*-e-a; after William Kent. Greenhouse and room palms.
BELMOREANA, bel-mor-e-*a*-na, after Belmore.
FORSTERIANA, fors-ter-e-*a*-na, after Forster.

**Kerria,** *ker*-re-a; named after William Kerr, plant collector, of Kew, who introduced K. japonica. Flowering shrubby climber.
JAPONICA, jap-*on*-ik-a, of Japan.

**Kirengeshoma,** kir-en-ge-*sho*-ma; the Japanese name for this herbaceous perennial.
PALMATA, pal-*ma*-ta, palmate, like a hand—the leaves.

**Kitaibelia,** kit-a-*be*-le-a; after Paul Kitaibel, an Austrian botanist. Border and woodland plants.
VITIFOLIA, vi-tif-*ol*-e-a, Vitis or vine-leaved.

**Kleinia,** *kly*-ne-a; after Dr. Klein, a German botanist. Greenhouse succulents.

**Kleinia** (*continued*)
ARTICULATA, ar-tik-ul-*a*-ta, jointed.
GALPINII, *gal*-pin-ei, after Galpin.
REPENS, *re*-penz, creeping and rooting.

**Kniphofia**, nif-*of*-e-a; after Johan Hieronymus Kniphof, a German professor of medicine. Herbaceous perennials.
ALOIDES, al-*oy*-deez, aloe-like.
BURCHELLII, *bur*-chel-lei, after Burchell.
CAULESCENS, kaw-*les*-senz, long-stemmed.
MACOWANII, mak-*ow*-an-ei, after Macowan.
NELSONII, *nel*-son-ei, after Nelson.
NORTHIÆ, *nor*-the-e, after Miss North.
RUFA, *roo*-fa, reddish.
TUCKII, *tuck*-ei, after Tuck.
TYSONII, *ti*-so-nei, after Tyson.
UVARIA, u-*var*-e-a, old generic name, meaning clustered. The Red-hot Poker.

**Kochia**, *kok*-e-a; after W. D. J. Koch, a German botanist. Foliage annual.
SCOPARIA, sko-*par*-e-a, broom-like.
TRICHOPHYLLA, trik-of-*il*-la, hair-leaved. Summer Cypress.

**Koelreuteria**, kol-roy-*teer*-e-a; after Köelreuter, a German botanist. Deciduous flowering tree.
PANICULATA, pan-ik-ul-*a*-ta, flowers in panicles.

**Koniga**, *kone*-ig-a; after Charles Kœnig Also spelled Kœniga. Hardy annual.
MARITIMA, mar-*it*-im-a, of sea coasts. The Sweet Alyssum.

**Laburnum**, la-*bur*-num; the old Latin name for the tree.
ALPINUM, al-*pine*-um, alpine.
VULGARE, vul-*gar*-e, common.

**Lachenalia**, lak-en-*a*-le-a; named after M. de Lachenal, a botanical author. Greenhouse bulbs.
AUREA, *aw*-re-a, golden.
NELSONII, nel-*so*-nei, after Nelson.
PENDULA, *pen*-du-la, hanging down.
TRICOLOR, *trik*-ol-or, three-coloured.

**Lactuca**, lak-*tu*-ka, from L. *lac*, milk, in reference to the white juice. Perennials and salad vegetable.
BOURGÆI, *boor*-je-i, after Bourgeau, plant collector.
PLUMIERI, ploo-me-*air*-i, after Plumier, a French botanist.
SATIVA, *sat*-iv-a, cultivated. The Lettuce.

**Lælia**, *la*-le-a; from Lælia, the name of a vestal virgin. Tropical orchids.
ALBIDA, *al*-bid-a, whitish.
ANCEPS, *an*-seps, having two edges.

**Lælia** (*continued*)
AUTUMNALIS, aw-tum-*na*-lis, autumn flowering.
CINNABARINA, kin-nab-ar-*e*-na, cinnabar; from Gr. *kinnabari*, a Persian dye.
CRISPA, *kris*-pa, curled.
ELEGANS, *el*-e-ganz, elegant.
PUMILA, *pu*-mil-a, dwarf.
PURPURATA, pur-pur-*a*-ta, purple.

**Læliocattleya**, *la*-le-o-*kat*-le-a; name for hybrids between Lælias and Cattleyas.

**Lagenaria**, lag-en-*ar*-e-a; from L. *lagena*, a bottle, referring to the shape of the fruit. Tender annuals.
VULGARIS, vul-*gar*-is, common.

**Lagerstrœmia**, la-ger-*stro*-me-a; after Magnus Lagerstrœm, Swedish friend of Linnæus. Shrubs or trees.
INDICA, *in*-dik-a, of India.

**Lagurus**, lag-*u*-rus; from Gr. *lagos*, a hare, and *oura*, a tail, alluding to the tail-like infloresence. Annual ornamental grass.
OVATUS, o-*va*-tus, oval—the infloresence. The Hare's-tail grass.

**Lamarckia**, lam-*ark*-e-a; after J. B. Lamarck, a French naturalist. Annual ornamental grass.
AUREA, *aw*-re-a, golden—the infloresence.

**Lamium**, *la*-me-um; from Gr. *laimos*, the throat, alluding to the throat-like appearance of the blossoms. Herbaceous perennials.
GALEOBDOLON, ga-le-*ob*-do-lon, old generic name, meaning a weasel and a bad smell.
MACULATUM, mak-ul-*a*-tum, spotted—the foliage with silvery stripe.

**Lampranthus**, lam-*pran*-thus; from Gr. *lampros*, brilliant, and *anthos*, a flower—the showy flowers. Greenhouse succulents.
AUREUM, *aw*-re-um, golden.
BLANDUM, *bland*-um, mild.
BROWNII, *brown*-ei, after Brown.
COCCINEUM, kok-*sin*-e-um, scarlet.
ROSEUM, *ro*-ze-um, rose-coloured.
SPECTABILE, spek-*tab*-il-e, showy.
VIOLACEUM, vi-o-*la*-ce-um, violet.

**Lantana**, lan-*ta*-na; an ancient name for Viburnum, the foliage of the two shrubs being similar. Greenhouse flowering shrubby plant.
CAMARA, ka-*mar*-a, after Camara. Florists' varieties have displaced species.

59

**Lapageria,** lap-a-*jeer*-e-a; commemorating Josephine Lapagerie, the wife of Napoleon Bonaparte. Greenhouse climbers.

ROSEA, *ro*-ze-a, rose-coloured.

**Lapeyrousia,** lap-a-*roo*-se-a; after de Lapeyrouse, French navigator. Hardy bulbous flowering plants.

CRUENTA, krew-*en*-ta, blood-coloured or crimson.

**Lardizabala,** lar-diz-*ab*-a-la; after Señor Michael Lardizabala y Uribe, a Spanish naturalist. Hardy flowering climber.

BITERNATA, *bit*-er-*na*-ta, twice ternate—the leaves.

**Larix,** *lar*-iks; old Latin name for the Larch, the English name being derived from it.

EUROPÆA, u-*ro*-pe-a, European. The Common Larch.

LEPTOLEPIS, lep-to-*le*-pis, slender-scaled —the cones.

**Lasiandra,** las-e-*an*-dra; from Gr. *lasios*, woolly, and *aner*, an anther, alluding to the woolly stamens. Greenhouse flowering shrubs.

MACRANTHA, mak-*ranth*-a, large-flowered.

**Lasthenia,** las-*the*-ne-a; said to be the name of a girl pupil of Plato. Hardy annuals.

GLABRATA, glab-*ra*-ta, smooth—the leaves.

**Lastrea,** *las*-tre-a; named after a French botanist, De Lastre. Hardy ferns.

ÆMULA, *em*-u-la, to rival, this fern rivalling others in possessing fragrance. The Hay-scented Buckler Fern.

CRISTATA, kris-*ta*-ta, crested. In this case word refers to the fringed margins of the fronds.

FILIX-MAS, *fil*-iks-mas, male fern, it being of tall and robust growth. The Male Fern.

MONTANA, mon-*ta*-na, of mountains.

SPINULOSA, spi-nul-*o*-sa, spiny—the margins of lobes and pinnules.

THELYPTERIS, thel-*ip*-ter-is, she or lady fern. The Marsh Fern.

**Latania,** lat-*a*-ne-a; after Latanier, the native name. Greenhouse palms.

BORBONICA, bor-*bo*-nik-a, of Isle of Bourbon.

**Lathyrus,** *lath*-e-rus; ancient Greek name for some leguminous plant. Annuals and perennials.

GRANDIFLORUS, gran-dif-*lo*-rus, large-flowered.

LATIFOLIUS, lat-if-*o*-le-us, broad-leaved. The Everlasting Pea.

ODORATUS, od-or-*a*-tus, fragrant. The Sweet Pea.

MAGELLANICUS, maj-el-*lan*-ik-us, from Straits of Magellan. Lord Anson's Pea.

TINGITANUS, ting-e-*ta*-nus, of Tangier. The Tangier Pea.

**Laurus,** *law*-rus, old Latin name for a Bay tree, the true "Laurel" of the ancients, perhaps derived from Celtic *laur*, green. Evergreen shrub.

NOBILIS, *no*-bil-is, noble. The Bay Tree.

**Lavandula,** lav-*an*-du-la; said to be derived from L. *lavo*, to wash, the Romans and Greeks having used Lavender in their baths. Sub-shrub.

SPICA, *spi*-ka, a spike, the shape of the inflorescence, the Spike Lavender of Spike-oil Plant.

VERA, *ve*-ra, true, alluding to its once having been regarded merely as a form of above and not a true species, or to the fact that it is the true old English Lavender of commerce.

**Lavatera,** la-*vat*-er-a; after J. K. Lavater, a Swiss naturalist of seventeenth century. Biennials and perennials; some sub-shrubby.

ARBOREA, ar-*bor*-e-a, tree-like.

OLBIA, *ol*-be-a, Olbia, old name for Hyères (S. France), where this plant abounds; also means rich.

TRIMESTRIS, trim-*es*-tris, maturing in three months.

**Layia,** *la*-e-a; after G. T. Lay, a naturalist and explorer. Hardy annuals.

ELEGANS, *el*-e-ganz, elegant.

PLATYGLOSSA, plat-e-*glos*-sa, broad-tongued.

**Ledum,** *le*-dum; probably from Gr. *ledos*, a woollen cloth, in reference to the woolly underparts of the leaves. Shrubs.

LATIFOLIUM, lat-if-*o*-le-um, broad-leaved.

**Leea,** *lee*-a; after James Lee, a noted Hammersmith nurseryman, 1715-1795. Tropical foliage shrubs.

AMABILIS, am-*a*-bil-is, beautiful.

**Leiophyllum,** li-o-*fil*-lum; from Gr. *leios*, smooth, and *phyllon*, a leaf, the foliage being glossy. Shrub.

BUXIFOLIUM, buks-if-*ol*-e-um, Buxus(box)-leaved.

**Lemaireocereus,** le-*mar*-o-*se*-re-us; after Charles Lemaire, a French cactus specialist. and *cereus*. Greenhouse cacti.

CANDELABRUM, kan-de-*la*-brum.

HYSTRIX, *his*-trix, like a porcupine.

**Lemna,** *lem*-na; from Gr. *lepis*, a scale, the form of the plants. Floating aquatics. The Duck weed.

MINOR, *mi*-nor, smaller.

TRISULCA, *tris*-ul-ka, in threes—the appearance of leafy growths.

**Leonotis,** le-on-*o*-tis; from Gr. *leon*, a lion, and *ous*, an ear, the flower having a fancied resemblance to a lion's ear. Greenhouse flowering shrub.

LEONURUS, le-o-*nu*-rus, lion's-tail—the flower spike.

**Leontopodium,** le-on-to-*po*-de-um; from Gr. *leon*, a lion, and *pous*, a foot, the flowers and leaves being supposed to resemble a lion's paw. Hardy perennial.

ALPINUM, al-*pine*-um, alpine. The Edelweiss.

**Leonurus,** le-on-*u*-rus; from Gr. *leon*, a lion, and *oura*, a tail, the tufted flowerhead suggesting a lion's tail. Herbaceous perennials.

CARDIACA, kar-*di*-a-ka, of the heart—ancient medicine.

**Lepidium,** lep-*id*-e-um; from Gr. *lepis*, a scale, the shape of the pods. The garden or salad cress.

SATIVUM, *sat*-iv-um, cultivated.

**Lepismium,** lep-*is*-me-um; from Gr. *lepis*, a scale, the small scales attached to the areoles. Greenhouse cacti.

COMMUNE, kom-*mu*-ne, communal or common.

**Leptosiphon,** lep-*tos*-if-on; from Gr. *leptos*, slender, and *siphon*, a tube, referring to the flowers of some species. Annuals and rock plants.

ANDROSACEUS, an-dro-*sa*-se-us, Androsace-like.

DENSIFLORUS, den-sif-*lo*-rus, the flowers clustered.

**Leptospermum,** lep-to-*sper*-mum; from Gr. *leptos*, slender, and *sperma*, a seed, the latter being very thin, almost threadlike. Shrubs.

SCOPARIUM, sko-*par*-e-um, broom-like (Cytisus scoparius).

**Leptosyne,** lep-*tos*-in-e; from Gr. *leptos*, slender, mainly annuals of slender growth.

DOUGLASII, dug-*las*-ei, after Douglas.

MARITIMA, mar-*it*-im-a, maritime.

STILLMANNII, *stil*-man-ei, after Stillmann.

**Leschenaultia,** les-ken-*ault*-e-a; after M. Leschenault, a French botanist. Greenhouse flowering shrubs.

BILOBA, *bil*-o-ba, two-lobed—the flowers.

FORMOSA, for-*mo*-sa, handsome.

**Lespedeza,** les-pe-*de*-za; after M. Lespedez, a patron of botany. Hardy flowering shrubs.

BICOLOR, *bik*-o-lor, two-coloured.

CAPITATA, kap-it-*a*-ta, flowers in a head.

**Leuchtenbergia,** look-ten-*ber*-ge-a; after Eugene de Bauharnaic, Duke of Leuchtenberg, French statesman. Greenhouse cacti.

PRINCIPIS, prin-*sip*-is, noble, princely.

**Leucojum,** lew-*ko*-e-um; from Gr. *leukos*, white, and *ion*, a violet, referring to the colour and possibly the fragrance of the flowers. Hardy bulbs. The Snowflake.

ÆSTIVUM, *es*-tiv-um, summer—season of flowering.

VERNUM, *ver*-num, spring—season of flowering.

**Leucothoe,** lew-*ko*-tho-e; said to be called after Leucothea, daughter of a Babylonian king (Grecian mythology), who, on being buried alive by her father, was transformed into a shrub by Apollo. Shrubs.

AXILLARIS, aks-il-*lar*-is, the flowers in axillary racemes.

CATESBÆI, *kates*-be-i, after Catesby, a naturalist of Carolina.

DAVISIÆ, da-*vis*-e-e, after a Miss N. J. Davis, who made the second discovery of the species.

**Lewisia,** lew-*is*-e-a; after a Captain Lewis —a traveller. Rock plants.

HOWELLII, *how*-el-ei, after Howell.

REDIVIVA, red-iv-*i*-va, reviving, the plant suddenly flowering, although apparently lifeless.

TWEEDYI, *tweed*-e-i, after Tweedy, a botanical collector.

**Leycesteria,** lay-ses-*teer*-e-a; commemorating W. Leycester, once Chief Justice in Bengal, L. formosa being a Himalayan shrub.

FORMOSA, for-*mo*-sa, handsome.

**Liatris,** li-*a*-tris; derivation unknown. Herbaceous perennials.

ELEGANS, *el*-e-ganz, elegant. The Blazing Star.

PYCNOSTACHYA, pik-nos-*tak*-e-a, densely spiked.

**Libertia,** li-*ber*-te-a; after M. A. Libert, a Belgian lady botanist. Half-hardy bulbs.

FORMOSA, for-*mo*-sa, handsome.

GRANDIFLORA, gran-dif-*lo*-ra, large-flowered.

IXIOIDES, iks-e-*oy*-dez, ixia-like.

**Libocedrus,** lib-o-*se*-drus; from Gr. *libanos*, the tree which yields frankincense, and *kedros*, a cedar. Evergreen conifers.

DECURRENS, de-*kur*-rens, running-down —the lower parts of leaves clasping the branches. The Incense Cedar.

**Libonia,** le-*bo*-ne-a; after M. Libon. Greenhouse shrubby flowering plants.

FLORIBUNDA, flor-ib-*un*-da, free flowering.

PENRHOSIENSIS, pen-rhos-e-*en*-sis, of Penrhos Hall, Wales.

**Ligustrum,** li-*gus*-trum; Latin name for Privet, possibly from L. *ligo*, to bind, the twigs having been used for tying. Shrubs.

CORIACEUM, kor-i-*a*-se-um, leathery, the leaves.

JAPONICUM, jap-*on*-ik-um, of Japan.

OVALIFOLIUM, o-val-if-*o*-le-um, oval-leaved. Oval-leaved Privet.

VULGARE, vul-*gar*-e, common. Common Privet.

**Lilium,** *lil*-e-um (classical, *le*-le-um); Latin name for Lily, and common to almost all European languages.

AURATUM, aw-*ra*-tum, golden-rayed.

CANDIDUM, *kan*-did-um, white. The White Lily.

CHALCEDONICUM, kal-se-*don*-ik-um, of Chalcedonia, Greece.

CROCEUM, *kro*-ke-um, saffron-coloured. The Orange Lily.

ELEGANS, *el*-e-ganz, elegant.

HANSONII, *han*-so-nei, after Hanson.

HENRYI, *hen*-re-i, after Dr. A. Henry, its discoverer.

LANCIFOLIUM, lan-sif-*ol*-e-um, lance-shaped leaves.

LONGIFLORUM, long-if-*lo*-rum, long-flowered.

MARTAGON, *mar*-ta-gon, old name of obscure origin.

MONADELPHUM, mon-a-*del*-fum, monadelphous, having the stamens united.

**Lilium** (*continued*)

PARDALINUM, par-da-*le*-num, panther-spotted.

PHILIPPENSE, fil-ip-*pen*-se, of Philippine Islands.

POMPONIUM, pom-*po*-ne-um, of much splendour.

REGALE, re-*ga*-le, royal, alluding to the magnificent flowers.

SPECIOSUM, spes-e-*o*-sum, showy.

SUPERBUM, su-*per*-bum, superb.

TESTACEUM, tes-*ta*-se-um, pale brown or apricot—colour of flowers.

TIGRINUM, tig-*re*-num, tiger-like. The Tiger Lily.

WILLMOTTIÆ, wil-*mot*-tee, after Miss Willmott.

**Limnanthemum,** lim-*nan*-the-mum; from Gr. *limne*, marsh or pool, and *anthemon*, a blossom, a reference to the plants being marsh or aquatic. Greenhouse and hardy aquatics.

HUMBOLDTIANUM, hum-*bolt*-e-an-um, after Humboldt, the traveller.

NYMPHÆOIDES, nim-fe-*oy*-dez, nymphæa-like.

PELTATUM, pel-*ta*-tum, shield-shaped— the leaves.

**Limnanthes,** lim-*nan*-thes; from Gr. *limne*, a marsh, and *anthos*, a flower, some of these annuals inhabiting moist places.

DOUGLASII, dug-*las*-ei, after Douglas, the plant collector.

**Limnobium,** lim-*no*-be-um; from Gr. *limne*, a marsh, and *bios*, life—living in marshy pools. Greenhouse submerged aquatic.

BOGOTENSIS, bo-go-*ten*-sis, from Bogota.

STOLONIFERUM, sto-lon-*if*-er-um, runner-bearing—the creeping stems.

**Limnophila,** lim-*nof*-il-a; from Gr. *limne*, a pool, and *philos*, loving—in reference to place of growth. Submerged aquatic.

GRATIOLOIDES, grat-e-ol-*oy*-dez, Gratiola-like.

SESSILIFLORA, ses-sil-if-*lo*-ra, flowers without footstalk.

**Limonium,** lim-*on*-e-um; from Gr. *limne*, a marsh, the plants growing in salt marshes. Greenhouse and hardy perennials.

BONDUELLII, bon-du-*el*-lei, after Bonduell.

EXIMIUM, eks-*im*-e-um, excellent.

LATIFOLIUM, lat-if-*ol*-e-um, broad-leaved.

PROFUSUM, pro-*fu*-sum, profuse.

SINUATA, sin-u-*a*-ta, scalloped-leaved.

**Limonium** (*continued*)
SUWOROWII, su-wor-*o*-ei, after Suworow.
VULGARE, vul-*gar*-e, common. The Sea
    Lavender.

**Linaria,** lin-*ar*-e-a; from L. *linum,* flax,
which some species resemble in growth.
Rock plants and annuals.
ÆQUITRILOBA, e-kwit-ril-*o*-ba, three equal
    lobes—the leaves.
ALPINA, al-*pine*-a, alpine.
BIPARTITA, bip-ar-*te*-ta, two-parted, pre-
    sumably the lip.
CYMBALARIA, sim-bal-*ar*-e-a, from Gr.
    *kymbalon,* a cymbal, the shape of the
    leaves. The Ivy-leaved Toadflax.
DALMATICA, dal-*mat*-e-*ka,* of Dalmatia.
HEPATICÆFOLIA, hep-at-ik-e-*fo*-le-a, hep-
    atica-leaved.
MACEDONICA, mas-e-*don*-ik-a, of Mace-
    donia.
MAROCCANA, mar-ok-*ka*-na, of Morocco.
PURPUREA, pur-*pur*-e-a, purple-flowered.
RETICULATA, re-tik-ul-*a*-ta, netted or
    striped.
TRISTIS, *tris*-tis, sad—the colour (brown).
VULGARIS, vul-*gar*-is, common. The
    Common Toadflax.

**Linnæa,** lin-*ne*-a; named after Linnæus,
this woodland plant, L. borealis, being
a great favourite of his.
BOREALIS, bor-e-*a*-lis, northern.

**Linum,** *li*-num, the Latin name for
flax (Gr. *linon*). Herbaceous, rock plants
and annuals.
ALPINUM, al-*pine*-um, alpine.
ARBOREUM, ar-*bor*-e-um, tree-like.
COCCINEUM, kok-*sin*-e-um, scarlet.
FLAVUM, *fla*-vum, yellow.
GRANDIFLORUM, gran-dif-*lo*-rum, large-
    flowered.
MONOGYNUM, mon-*o*-jin-um, having a
    single style.
NARBONENSE, nar-bon-*en*-se, of Nar-
    bonne.
PERENNE, per-*en*-ne, perennial.
SALSOLOIDES, sal-sol-*oy*-dez, like Salsola
    (Saltwort).
USITATISSIMUM, u-sit-a-*tis*-sim-um, most
    commonly used. The Flax Plant.

**Lippia,** *lip*-pe-a; after Augustus Lippi, a
French traveller. Shrubs and rock plants.
CITRIODORA, sit-re-od-*or*-a, lemon-scented
    —the foliage. The Lemon-scented
    Verbena.

**Liquidambar,** lik-wid-*am*-bar; from L.
*liquidus,* liquid, and *ambar,* amber, in
reference to the gum (storax) yielded by
some species. Trees.
STYRACIFLUA, ster-ak-*if*-lu-a, storax-flow-
    ing, hence popular name, Sweet Gum.

**Liriodendron,** lir-e-o-*den*-dron; from Gr.
*lirion,* a lily, and *dendron,* a tree. Trees.
TULIPIFERA, tew-lip-*if*-er-a, tulip-bearing.
    The Tulip Tree.

**Lithops,** *lith*-ops; from Gr. *lithos,* a
stone, and *ops,* the face, on account of
the resemblance to the stones (pebbles)
among which the plants grow. Greenhouse succulents. The Pebble Plants.
BELLA, *bel*-la, beautiful or pretty.
FULVICEPS, *ful*-ve-seps, tawny or rust
    coloured—the plant.
LESLIEI, *les*-le-i, after Leslie.
TURBINIFORMIS, ter-bin-if-*or*-mis, top-
    or cone-shaped.

**Lithospermum,** lith-o-*sper*-mum; from
Gr. *lithos,* a stone, and *sperma,* a seed,
the latter being extremely hard. Sub-
shrubby rock plants.
GRAMINIFOLIUM, gram-in-if-*ol*-e-um,
    grass-leaved.
PROSTRATUM, pros-*tra*-tum, prostrate.
PURPUREO-CŒRULEA, pur-*pew*-re-o-se-*ru*-
    le-a, purple and blue—the flowers.

**Littorella,** lit-tor-*el*-la; from L. *littus,* the
shore—inhabits sandy pools. Hardy
aquatic.
LACUSTRIS, lak-*us*-tris, found in lakes.
UNIFLORA, u-nif-*lo*-ra, one-flowered—
    the flowers on single stems.

**Livistona,** liv-is-*to*-na; after Patrick
Murray of Livingstone, near Edin-
burgh. Greenhouse palms.
SINENSIS, sin-*en*-sis, of China.
HUMILIS, *hum*-il-is, small or humble—the
    size of plant.
ROTUNDIFOLIA, ro-tun-dif-*ol*-e-a, round
    leaved.

**Loasa,** lo-*a*-sa; native S. American name,
meaning and derivation not known.
Greenhouse annual climbing plants.
LATERITIA, lat-er-*it*-e-a, brick-red—the
    colour of the flowers.

**Lobelia,** lo-*be*-le-a; after M. Matthias de
Lobel, a Fleming, physician to James I,
traveller, plant collector, and botanical
author. Hardy and half-hardy her-
baceous perennials and annuals. Many
garden varieties.
CARDINALIS, kar-din-*a*-lis, deep scarlet.
ERINUS, e-*ri*-nus, old generic name. The
    bedding lobelia.
FULGENS, *ful*-jenz, glowing—the brilliant
    flowers.
SYPHYLITICA, sif-il-*it*-ik-a, alluding to the
    disease, for which the plant was once
    used as a remedy.

**Lobelia** (*continued*)

TENUIOR, *ten*-u-e-or, more slender.
TUPA, *tew*-pa, the old generic name of native Chilean origin.

**Lobivia**, lob-*iv*-e-a; an anagram of Bolivia, where the species are found. Greenhouse cacti.

CÆSPITOSA, ses-pit-*o*-sa, growing in tufts.
FEROX, *fer*-oks, fierce—the spines.
GRANDIS, *grand*-is, grand or large.
GRANDIFLORA, gran-dif-*lo*-ra, large flowered.
HAAGEANA, haag-e-*a*-na, after Haage.
PENTLANDII, *pent*-land-ei, after Pentland.
REBUTIOIDES, re-*but*-e-oy-dez, Rebutia-like.

**Loiseleuria**, loys-el-*eur*-e-a; after Loiseleur Deslongchamps, a French botanist. Trailing flowering shrubs.

PROCUMBENS, pro-*kum*-benz, lying down.

**Lomaria**, lo-*mar*-e-a; from Gr. *loma*, a margin, in reference to the position of the sori. Greenhouse and hardy ferns.

ALPINA, al-*pine*-a, alpine.
GIBBA, *gib*-ba, gibbous or humped.
SPICANT, *spi*-kant, spiked—the pointed fertile fronds.

**Lomatia**, lo-*ma*-te-a; from Gr. *loma*, a border, referring to the winged edge to the seeds. Greenhouse foliage shrubs.

ELEGANTISSIMA, el-eg-an-*tis*-sim-a, most elegant.
LONGIFOLIA, long-if-*ol*-e-a, long-leaved.

**Lonicera**, lon-*is*-er-a; named after Adam Lonicer, a German naturalist of sixteenth century. Climbers and shrubs.

AUREO-RETICULATA, *aw*-re-o-ret-ik-ul-*a*-ta, golden veined; a variety of L. japonica.
CAPRIFOLIUM, kap-rif-*o*-le-um, herbalist name, a plant which climbs like a goat.
FRAGRANTISSIMA, fra-gran-*tis*-sim-a, most fragrant.
NITIDA, *nit*-id-a, shining—the glossy leaves.
PERICLYMENUM, per-ik-*lim*-en-um, to twine around. The Woodbine or Honeysuckle.
PILEATA, pil-e-*a*-ta, having a cap, the berry being topped by a curious outgrowth of the calyx.
SEMPERVIRENS, sem-per-*veer*-enz, always green.
SYRINGANTHA, syr-ing-*an*-tha, the flowers resembling Syringa (lilac).
XYLOSTEUM, zy-*los*-te-um, Xylosteum—a disused generic name from Gr. *xylon*, wood—the woody stems.

**Lophocereus**, *lof*-o-se-*re*-us; from Gr. *lophos*, a crest, and *cereus*, referring to the bristly top of the stem when flowering. Greenhouse cacti.

SCHOTTII, *shot*-tei, after Schott. The variety monstrosa is spineless and irregular in shape and looks like a sculpture in green jade.

**Lophophora**, lof-*of*-or-a; from Gr. *lophos*, a crest, and *phoreo*, I bear, the reference is to the hairs borne at the aeroles. Greenhouse cacti.

LEWINII, *lew*-in-ei, after Lewin.
WILLIAMSII, *will*-yams-ei, after Williams.

**Loropetalum**, lo-rop-*et*-al-um; from Gr. *loron*, a thong, and *petalon*, a petal, in allusion to the thong-shaped petals. Deciduous flowering shrubs.

SINENSIS, sin-*en*-sis, of China.

**Lotus**, *lo*-tus; old name adopted by Greek naturalists for a trefoil-like plant. Greenhouse and hardy perennials.

CORNICULATUS, kor-nik-ul-*a*-tus, a little horn, the shape of the flower. Bird's-foot Trefoil.
PELIORHYNCHUS, pel-e-or-*in*-kus, stork's beak.

**Luculia**, lu-*ku*-le-a; probably from Luculi Swa, the native Nepalese name. Greenhouse evergreen flowering shrub.

GRATISSIMA, gra-*tis*-sim-a, most welcome or very grateful—the fragrance.
PINCEANA, *pin*-se-a-na, after Pince.

**Ludwigia**, lud-*vig*-e-a; after Kristian Gottlieb Ludwig, botanist and professor of medicine. Bog plants or submerged aquatics.

MULERTTII, mul-*ert*-tei, after Mulertt.
PALUSTRIS, pal-*us*-tris, of marshes.

**Lunaria**, loon-*air*-e-a; from L. *luna*, the moon, alluding to the round and silvery seed vessels. Hardy biennials.

BIENNIS, bi-*en*-nis, of biennial duration. The Honesty.

**Lupinus**, lu-*pe*-nus (or lu-*py*-nus); from L. *lupus*, a wolf (destroyer), some species devastating land by their abundance. Annuals and herbaceous perennials.

ARBOREUS, ar-*bor*-e-us, tree-like. The Tree Lupin.
HARTWEGII, hart-*ve*-gei, after Hartweg.
MUTABILIS, mu-*ta*-bil-is, variable in form or colour.
NOOTKATENSIS, noot-kat-*en*-sis, of Nootka Sound.

**Lupinus** (*continued*)

PAYNEI, *pa*-ne-i, after Th. Payne of California, who discovered it.

POLYPHYLLUS, pol-if-*il*-lus, many-leaved. The Common Lupin.

PUBESCENS, pu-*bes*-senz, downy.

SUBCARNOSUS, sub-kar-*no*-sus, somewhat fleshy.

**Lycaste,** ly-*kas*-tee; after Lycaste, a Sicilian beauty. Greenhouse orchids.

AROMATICA, ar-o-*mat*-ik-a, aromatic.

DEPPEI, *dep*-pe-i, after Deppe.

SKINNERI, *skin*-ner-i, after Skinner.

**Lychnis,** *lik*-nis; from Gr. *lychnos*, a lamp, in reference to the brilliantly coloured flowers. Herbaceous border and rock plants.

ALPINA, al-*pine*-a, alpine.

CHALCEDONICA, kal-se-*don*-ik-a, of Chalcedonia.

CŒLI-ROSA, kee-le-*ro*-za, Rose of Heaven.

CORONARIA, kor-o-*na*-re-a, crowned.

DIOICA, di-*o*-ik-a, literally, two houses, *i.e.*, a diœcious plant.

DIURNA, di-*ur*-na, day-flowering. Double kind chiefly cultivated.

FLOS-CUCULI, flos-*kuk*-ul-e, flower of the cuckoo, *i.e.*, blooming in cuckoo-time.

FLOS-JOVIS, flos-*jo*-vis, flower of Jove, name of ancient origin.

HAAGEANA, haag-e-*a*-na, after Haage.

VISCARIA, vis-*kar*-e-a, sticky, the stems gummy.

VESPERTINA, ves-per-*tee*-na, evening flowering. Double form cultivated.

**Lycium,** *ly*-se-um; said to be from Gr. *lykion*, a plant of Lycia, in Asia Minor. Shrubs.

BARBARUM, *bar*-bar-um, foreign.

CHINENSE, tshi-*nen*-se, of China.

**Lycopersicum,** lik-op-*er*-sik-um (or ly-co-*per*-se-kum); from Gr. *lykos*, a wolf, and *persicon*, a peach, probably in reference to supposed poisonous qualities. Greenhouse annual fruit-bearing plant.

ESCULENTUM, es-kul-*en*-tum, edible. The Tomato.

**Lycoris,** ly-*kor*-is; a mythological name. Greenhouse bulbs.

AUREA, *aw*-re-a, golden.

RADIATA, rad-e-*a*-ta, radiating from a centre.

SQUAMIGERA, skwam-*ig*-er-a, scale-bearing.

**Lygodium,** ly-*go*-de-um; from Gr. *lygodes*, flexible, referring to the twining stems. Greenhouse twining ferns.

**Lygodium** (*continued*)

JAPONICUM, jap-*on*-ik-um, of Japan.

PALMATUM, pal-*ma*-tum, hand-shaped.

SCANDENS, skan-denz, scandent or climbing.

**Lyonia,** ly-*o*-ne-a; after J. Lyon, an American plant collector. Dwarf peat shrubs.

LIGUSTRINA, lig-us-*trin*-a, resembling Ligustrum (privet).

**Lysichiton,** lis-e-*ki*-ton; from Gr. *lusis*, loosing, or freeing from, and *chiton*, a tunic, alluding to the wide-open spathe. Herbaceous perennials.

AMERICANUM, a-mer-ik-*a*-num, of America.

KAMTSCHATKENSE, kams-kat-*ken*-se, of Kamtschatka.

**Lysimachia,** lis-e-*mak*-e-a; probably from Gr. *luo*, to loose, and *mache*, strife, hence Loosestrife. Some authorities state that the genus is named after Lysimachus, King of Thrace (306 B.C.), who, it is said, was the first to discover the Loosestrife's supposed soothing properties. Herbaceous perennials.

CLETHROIDES, kleth-*roy*-des, clethra-like —the flowers.

NUMMULARIA, num-mul-*ar*-e-a, from L. *nummus*, a coin, the shape of the leaf. The Creeping Jenny.

PUNCTATA, punk-*ta*-ta, dotted—the flowers.

VERTICILLATA, ver-tis-il-*la*-ta, whorled— the foliage.

VULGARIS, vul-*gar*-is, common. The Yellow Loosestrife.

**Lythrum,** *lith*-rum, from Gr. *lythron*, blood, in allusion to the colour of the flowers. Herbaceous perennials.

ALATUM, al-*a*-tum, winged—the stalks.

SALICARIA, sal-ik-*ar*-e-a, willow-like—the leaves, or willow-herb-like—the flower spikes. The Purple Loosestrife.

VIRGATUM, ver-*ga*-tum, twiggy.

**Machærocereus,** mak-*a*-ro-se-*re*-us; from Gr. *machaira*, a dagger, and *cereus*, referring to the spines. Greenhouse cacti.

ERUCA, er-*oo*-ka, caterpillar-like—the prostrate stems. Known in America as the "Creeping Devil."

**Mackaya,** mak-*kay*-a; after Dr. J. F. Mackay, Keeper of Dublin University Botanic Garden. Greenhouse flowering shrub.

BELLA, *bel*-la, pretty.

**Magnolia,** mag-*no*-le-a; named by Linnæus in commemoration of Pierre Magnol, a professor of botany and medicine at Montpellier in sixteenth century. Shrubs and trees.

CONSPICUA, kon-*spik*-u-a, conspicuous, showy.

GRANDIFLORA, gran-dif-*lor*-a, large-flowered.

KOBUS, *ko*-bus, a Japanese name.

LENNEI, *len*-ne-i, after Lenne, once Royal gardener in Berlin.

OBOVATA, ob-ov-*a*-ta, obovate.

SOULANGEANA, soo-lan-je-*a*-na, after Soulange-Bodin, botanist, of Fromont (France).

STELLATA, stel-*la*-ta, starry—the flowers.

**Mahonia,** ma-*ho*-ne-a; after M'Mahon of North America. A group of Barberries of which Berberis aquifolium is typical, sometimes united with that genus.

**Maianthemum,** ma-*an*-the-mum; from Gr. *Maia,* the mother of Mercury (Gr. mythology), to whom the month of May was dedicated, and *anthemon,* a flower. Rock and woodland plants.

CONVALLARIA, con-val-*lar*-e-a, bot. name for Lily of the Valley, which this plant resembles.

**Malacocarpus,** mal-a-ko-*kar*-pus; from Gr. *malakos,* soft, and *karpos,* fruit, referring to the soft fleshy fruits. Greenhouse cacti.

CORYNOIDES, kor-e-*noy*-dez, club-like—the stigmas.

ERINACEUS, er-in-*a*-se-us, erinacea-like.

**Malcolmia,** mal-*ko*-me-a; after William Malcolm, nurseryman, botanist, and associate of the naturalist Ray. Hardy annuals.

MARITIMA, mar-*it*-im-a; maritime. The Virginian Stock.

**Malope,** *mal*-o-pe; old Gr. name for a kind of mallow, meaning soft or soothing, from the leaf texture or medicinal properties. Hardy annuals.

GRANDIFLORA, gran-dif-*lo*-ra, large-flowered.

TRIFIDA, *trif*-id-a, three-cleft—the leaves.

**Malva,** *mal*-va; L. from Gr. *malakos,* soft or soothing, probably alluding to an emollient yielded by the seeds. Annuals and herbaceous perennials.

ALCEA, *al*-se-a, old generic name.

CRISPA, *kris*-pa, crisp or curled at edges —the leaves.

MOSCHATA, mos-*ka*-ta, musk—the leaves slightly fragrant. The Musk Mallow.

**Malvastrum,** mal-*vas*-trum; from Malva and L. *aster,* a star. Rock plants and hardy perennials.

COCCINEUM, kok-*sin*-e-um, scarlet.

GILLIESII, gil-*leez*-ei, after Gillies.

LATERITIUM, lat-er-*it*-e um, brick-red.

**Mammillaria,** mam-mil-*lar*-e-a; from L. *mamma,* the breast, or *mamilla,* a nipple, in reference to the teat-like tubercles characteristic of many species. Greenhouse cacti.

APPLANATA, ap-pla-*na*-ta, flattened.

CHIONOCEPHALA, ki-on-o-*sef*-a-la, snowy-headed.

CIRRHIFERA, sir-*rif*-er-a, tendril-bearing.

DOLICHOCENTRA, dol-i-ko-*sen*-tra, long-spined.

ECHINATA, ek-in-*a*-ta, hedgehog-spined.

ELEGANS, *el*-e-ganz, elegant.

ELONGATA, e-long-*a*-ta, elongated.

GRACILIS, *gras*-il-is, slender.

LONGIMAMMA, long-e-*mam*-ma, long-teated.

MAGNIMAMMA, mag-ne-*mam*-ma, large-teated.

MICROTHELE, mi-kro-*the*-le, small-nippled.

NIVEA, *niv*-e-a, snowy.

PUSILLA, pu-*sil*-la, small.

SEMPERVIVI, sem-per-*vi*-ve, like a sempervivum.

STELLA-AURATA, *stel*-la-aw-*ra*-ta, golden star.

TETRACANTHA, tet-ra-*kan*-tha, four-spined.

UNCINATA, un-sin-*na*-ta, hooked.

VILLIFERA, vil-*lif*-er-a, bearing shaggy hair.

VIVIPARA, vi-*vip*-ar-a, viviparous; producing young stem plants.

**Mandevilla,** man-de-*vil*-la; after H. J. Mandeville, once British Minister at Buenos Ayres. Greenhouse evergreen climber.

SUAVEOLENS, swa - *ve* - o - lenz, sweet-scented.

**Mandragora,** man-dra-*gor*-a; from Gr. *mandragoras* (Mandrake), a herb possessing narcotic properties. Hardy perennials.

OFFICINARUM, of-fis-in-*ar*-um, of the shop, herbal.

**Manettia,** man-*et*-te-a; after Xavier Manetti, Prefect of Botanic Garden at Florence.

BICOLOR, bik-ol-or, two-coloured.

LUTEO-RUBRA, *loo*-te-o-*roob*-ra, yellow and red.

**Maranta,** mar-*an*-ta, after B. Maranti, an Italian botanist. Tropical foliage perennials.

**Maranta** (*continued*)

ARUNDINACEA, VARIEGATA, ar-un-din-*a*-se-a var-e-eg-*a*-ta, reed-like and variegated.

BELLULA, *bel*-u-la, little, pretty—neat, diminutive.

BICOLOR, *bik*-ol-or, two-coloured.

GRATIOSA, gra-ti-*o*-sa, favoured, the beautiful leaf-colour.

INSIGNIS, in-*sig*-nis, remarkable.

LEOPARDINA,lep-ar-*di*-na,leopard-spotted.

LINEATA, lin-e-*a*-ta, lined.

MASSANGEANA, mas-san-ge-*a*-na, after Massange.

MUSAICA, mew-*za*-ik-a, mosaic.

PICTA, *pik*-ta, painted.

POLITA, pol-*e*-ta, polished.

ROSEO-PICTA, *ro*-ze-o-*pik*-ta, rosy-painted.

WAGNERI, vag-*neer*-i, after Wagner, a German botanist.

WARSCEWICZI, var-skew-*ik*-zi, after Warscewicz, a Russian botanist.

ZEBRINA, ze-*bry*-na, zebra-striped.

**Margyricarpus,** mar-ge-re-*kar*-pus; from L. *margarita,* a pearl, and *karpos,* a berry. Half-hardy shrub.

SETOSUS, se-*to*-sus, bristly. The Pearl Berry.

**Marica,** *mar*-ik-a; from Gr. *maraino,* to flag, the flowers fading quickly. Greenhouse perennials.

BRACHYPUS, brak-*ip*-us, short-stalked.

CÆRULEA, se-*ru*-le-a, blue.

NORTHIANA, north-e-*a*-na, after Miss North.

**Marrubium,** mar-*roo*-be-um; believed to be the Hebrew name of one of the five bitter herbs eaten by the Jews at the Feast of the Passover, but Linnæus derived the word from Marrubium, a city of Latium, Italy. Aromatic herbs.

VULGARE, vul-*gar*-e, common. The Horehound.

**Marsdenia,** mars-*de*-ne-a; after W. Marsden, an author. Greenhouse and hardy shrubs.

ERECTA, e-*rek*-ta, upright.

FLAVESCENS, fla-*ves*-senz, becoming yellow.

MACULATA, mak-ul-*a*-ta, spotted—the leaves.

**Marshallia,** mar-*shal*-le-a; after H. Marshall, a botanical author. Half-hardy perennials.

LANCEOLATA, lan-se-o-*la*-ta, spear-shaped —the leaves.

**Marsilea,** mar-*se*-le-a; after Giovanni Marsigli, an Italian botanist. Cryptogamic aquatics.

QUADRIFOLIA, kwad-rif-*ol*-e-a, four leaves or four lobes to the leaves.

**Martynia,** mar-*tin*-e-a; after Dr. John Martyn, one-time professor of botany at Cambridge. Half-hardy annuals.

FRAGRANS, *fra*-granz, fragrant.

PROBOSCIDEA, pro-bos-*sid*-e-a, proboscis-like—the seed pods.

**Masdevallia,** maz-dev-*al*-le-a; after Joseph Masdevall, a Spanish botanist. Greenhouse orchids.

AMABILIS, am-*a*-bil-is, beautiful.

BELLA, *bel*-la, pretty.

CAUDATA, *kaw*-da-ta, tailed.

CHIMÆRA, *kim*-er-a, weird, fanciful—the flowers.

COCCINEA, kok-*sin*-e-a, scarlet.

HARRYANA, har-re-*a*-na, after Harry.

IGNEA, *ig*-ne-a, fiery—colour of flowers.

LINDENI, *lin*-den-e, after Linden.

TOVARENSIS, to-var-*en*-sis, of Tovar, Columbia.

VEITCHIANA, veech-e-*a*-na, of Messrs. Veitch, nurserymen of Chelsea.

**Matricaria,** mat-re-*kar*-e-a; from L. *matrix,* the womb—once used medicinally in feminine disorders. Hardy perennials.

EXIMIA, eks-*im*-e-a, choice.

INODORA PLENISSIMA, in-od-*o*-ra ple-*nis*-sim-a, very double, scentless; garden variety of Mayweed.

**Matthiola,** mat-*te*-o-la (also Mathiola, ma-*te*-o-la); after Piero Antonio Matthioli (or Peter Andrew Mathioli), Italian physician and botanist. Half-hardy annuals and biennials. Stock-gilliflower. Ten-week Stock, Brompton Stock, etc., of hybrid origin, have their parentage in this genus.

ANNUA, *an*-nu-a, of annual duration. The Ten-week Stock.

BICORNIS, bik-*or*-nis, two-horned—the seed pod. The Night-scented Stock.

INCANA, in-*ka*-na, hoary—the leaves. Brompton and allied Stocks.

SINUATA, sin-u-*a*-ta, wavy or scalloped— the leaves.

**Maurandya,** maw-*ran*-de-a; after Professor Maurandy of Cartagena. Half-hardy twiners. Sometimes written Maurandia.

BARCLAYANA, bar-kla-*a*-na, after Barclay.

**Maurandya** (*continued*)

SCANDENS, *skan*-denz, scandent or climbing.

**Maxillaria**, maks-il-*lar*-e-a; from L. *maxillæ*, the jaws of an insect, referring to the appearance of the column and lip. Greenhouse orchids.

GRANDIFLORA, gran-dif-*lo*-ra, large flowered.

HARRISONIÆ, har-ris-*o*-ne-e, after Mrs. Harrison.

PICTA, *pik*-ta, painted.

SANDERIANA, san-der-e-*a*-na, after Sander, nurseryman.

TENUIFOLIA, ten-u-if-*ol*-e-a, narrowleaved.

VENUSTA, ven-*us*-ta, charming or lovely.

**Mazus**, *ma*-zus; from Gr. *mazos*, a teat, in reference to the tubercles at the mouth of the flowers. Creeping rock plants.

PUMILIO, pew-*mil*-e-o, dwarf.

REPTANS, *rep*-tanz, creeping and rooting.

RUGOSUS, roo-*go*-sus, wrinkled—the leaves.

**Meconopsis**, mek-on-*op*-sis; from Gr. *mekon*, a poppy, and *opsis*, like. Biennials and perennials.

BAILEYI, *ba*-le-i, after Major Bailey, who first discovered it. The Blue Poppy.

BETONICIFOLIA, bet-on-ik-e-*fo*-le-a, betony (Betonica)-leaved.

CAMBRICA, *kam*-brik-a, of Wales. The Welsh Poppy.

GRANDIS, *gran*-dis, grand.

INTEGRIFOLIA, in-teg-rif-*ol*-e-a, leaves entire, not divided.

PRATTII, *prat*-ei, after Pratt, plant collector.

QUINTUPLINERVIA, kwin-tup-lin-*er*-ve-a, five-nerved—the leaves.

SIMPLICIFOLIA, sim-plis-if-*ol*-e-a, leaves simple, that is, not divided.

WALLICHII, wol-*lich*-ei, after Dr. Wallich.

**Medeola**, med-*e*-o-la; called after *Medea*, the Greek sorceress. Greenhouse twining perennials.

ASPARAGOIDES, as-par-ag-*oy*-dez, asparagus-like.

**Medicago**, med-ik-*a*-go; indirectly from Media, the country whence alfalfa was supposed to have been derived. Herbaceous perennials and fodder plants.

ECHINUS, ek-*in*-us, a hedgehog—the spiny seed pods.

FALCATA, fal-*ka*-ta, sickle-shaped—the pods.

SCUTELLATA, skew-tel-*la*-ta, a little shield.

**Medinilla**, me-din-*il*-la; after J. de Medinilla y Pineta, Spanish governor of the Marianne Islands. Tropical flowering shrubs.

CURTISII, ker-*tis*-ei, after Curtis.

MAGNIFICA, mag-*nif*-ik-a, magnificent.

**Megasea**, meg-*as*-e-a; from Gr. *megas*, great, the size of the plants, compared with other saxifrages. Hardy perennials.

CILIATA, sil-e-*a*-ta, ciliated or fringed with hairs.

CORDIFOLIA, kor-dif-*ol*-e-a, heart-shaped foliage.

CRASSIFOLIA, kras-sif-*ol*-e-a, thick-leaved.

STRACHEYI, *strak*-e-e, after Strachey.

**Melaleuca**, mel-a-*lew*-ka; from Gr. *melas*, black, and *leukos*, white, the colours of the old and new bark. Greenhouse shrubs.

ERICIFOLIA, er-ik-if-*o*-le-a, erica-leaved.

FULGENS, *ful*-jens, glowing.

LEUCADENDRON, loo-kad-*en*-dron, whiteleaved like Leucadendron.

SCABRA, *ska*-bra, rough—the leaves.

THYMIFOLIA, ty-mif-*ol*-e-a, thyme-leaved.

**Melianthus**, me-li-*an*-thus; from Gr. *mele*, honey, and *anthos*, a flower, the blooms of M. major yielding a honeylike juice. Half-hardy perennial.

COMOSUS, kom-*o*-sus, tufted, with hairs.

MAJOR, *ma*-jor, greater.

PECTINATA, pek-tin-*a*-ta, comb-like.

**Meliosma**, mel-i-*oz*-ma; from Gr. *mele*, honey, and *osme*, a smell, the scent of the flowers. Deciduous trees and shrubs.

MYRIANTHA, mir-e-*an*-tha, myriad-flowered.

VEITCHIORUM, veech-e-*or*-um, after Veitch, the famous Chelsea nursery firm.

**Melissa**, mel-*is*-sa; from Gr. *melissa*, a bee, presumably in allusion to the plants as honey producers.

OFFICINALIS, of-fis-in-*a*-lis, of the shop (herbal). The Common Balm. Variegated variety grown in gardens.

**Melittis**, mel-*it*-tis; same derivation as Melissa.

MELISSOPHYLLUM, mel-is-sof-*il*-lum, leaved like Melissa, hence Bastard Balm —the common name.

**Melocactus**, mel-o-*kak*-tus; from *melon*, the Melon, and *kaktos*, a spiny plant, so named by Theophrastus and descriptive of the Melon cactus, a globular species. Greenhouse cacti.

COMMUNIS, kom-*mu*-nis, common or social. The Melon Cactus.

**Menispermum**, me-nis-*per*-mum; from Gr. *mene*, the moon, and *sperma*, seed, the latter being crescent-shaped. Hardy climbing shrub.

CANADENSE, kan-a-*den*-se, Canadian. The Moonseed.

**Mentha**, *men*-tha; Gr. name of a nymph. Herbaceous perennials and carpeting plants.

AQUATICA, a-*kwat*-ik-a, growing in water. The Bergamot Mint.

GIBRALTARICA, gib-ral-*tar*-ik-a, of Gibraltar. The Gibraltar Mint.

PIPERITA, pi-per-*e*-ta, pepper. The Peppermint.

PULEGIUM, pu-*le*-je-um, from L. *pulex*, a flea, which the plant was supposed to eradicate. The Pennyroyal.

REQUIENII, re-*kwe*-nei, after Requien.

VIRIDIS, *ver*-id-is, green. The Lamb or Spear Mint.

**Mentzelia**, men-*ze*-le-a; named after Christian Mentzel, a German botanist; mainly annuals and biennials.

BARTONIOIDES, bar-to-ne-*oy*-deez, Bartonia-like.

LINDLEYI, *lind*-le-i, after Dr. Lindley, the botanist.

**Menyanthes**, men-e-*an*-thes; from Gr. *men*, a month, and *anthos*, a flower, the flowering period of the Bogbean, being supposed to last a month. Hardy aquatic.

CRISTA-GALLI, kris-ta-*gal*-le, cock's-crest.

NYMPHÆOIDES, nim-fe-*oy*-dez, nymphæa-like.

TRIFOLIATA, trif-o-le-*a*-ta, three-leaved. The Bogbean.

**Menziesia**, men-*ze*-ze-a; more correctly men-*e*-se-a (Scotch, ming-*is*-e-a), named after Archibald Menzies, surgeon-botanist on Vancouver's expedition, 1790-1795. Dwarf shrubs.

CÆRULEA, se-*ru*-le-a, blue.

EMPETRIFOLIA, em-pet-rif-*ol*-e-a, Empetrum-leaved.

POLIFOLIA, pol-if-*ol*-e-a, Polium-leaved —Teucrium Polium.

**Mertensia**, mer-*ten*-se-a; named after F. C. Mertens, professor of botany at Bremen. Hardy perennials.

ALPINA, al-*pine*-a, alpine.

CILIATA, sil-e-*a*-ta, fringed with hairs.

ECHIOIDES, ek-e-*oy*-dez, echium-like.

PRIMULOIDES, prim-ul-*oy*-des, primula-like.

SIBIRICA, si-*bir*-ik-a, of Siberia.

VIRGINICA, ver-*jin*-ik-a, of Virginia.

**Mesembryanthemum**, mes-em-bre-*an*-the-mum; from Gr. *mesembria*, midday, and *anthemon*, a flower, probably in reference to the blossoms opening best in sunshine. Mostly greenhouse succulents. This genus of many species, but with a general family likeness in the flowers, has been split by modern botanists into some ninety genera or sub-genera, the chief of which have been included in this volume. The subjoined are the best-known species in general cultivation. The Fig Marigold.

ACINACIFORME, as-in-as-if-*or*-me, scimitar-shaped—the leaves.

AURANTIACUM, aw-*ran*-te-ak-um, orange coloured.

AUREUM, *aw*-re-um, golden.

BARBATUM, bar-*ba*-tum, having hooked hairs.

BLANDUM, *bland*-um, pleasing.

BOLUSII, *bo*-lus-ei, after Bolus.

BROWNII, *brown*-ei, after Brown.

COCCINEUM, kok-*sin*-e-um, scarlet.

COOPERI, *koop*-er-e, after Cooper.

CORDIFOLIUM VARIEGATUM, kor-dif-*ol*-e-um var-e-eg-*a*-tum, variegated heart-shaped—the leaves.

CRINIFLORUM, krin-if-*lo*-rum, hair-like —the petals.

CRYSTALLINUM, kris-tal-*le*-num, crystalline—the foliage. The Ice Plant.

CYMBIFOLIUM, kim (or sim)-bif-*ol*-e-um, boat-shaped leaves.

DELTOIDES, del-*toy*-dez, delta-shaped— the leaves.

DENSUM, *den*-sum, close, dense.

ECHINATUM, ek-in-*a*-tum, hedgehog-like —the spines.

EDULE, ed-*u*-le, eatable—the fruits. The Hottentot Fig.

INCLAUDENS, in-*klaw*-denz, not closing —the flowers remaining expanded.

POLYANTHUM, pol-e-*anth*-um, many flowered.

POMERIDIANUM, po-mer-*id*-e-*a*-num, postmeridian, *i.e.*, flowers opening after midday.

PUGIONIFORME, pu-ge-e-o-nif-*or*-me, dagger shaped.

PYROPÆUM, py-ro-*pe*-um, flame-coloured.

ROSEUM, *ro*-ze-um, rosy.

SPECTABILE, spek-*tab*-il-e, showy.

TIGRINUM, tig-*re*-num, tiger-like—the spiny foliage.

TRICOLORUM, trik-ol-*or*-rm, three-coloured.

UNCINATUM, un-sin-*a*-tum, hooked at the end—the leaves.

VIOLACEUM, vi-ol-*a*-ce-um, violet-coloured.

**Mespilus**, *mes*-pil-us; from Gr. *mesos*, half, and *pilos*, a ball, referring to the half-ball shape of the fruits. Fruit bearing tree.

GERMANICA, jer-ma*n*-ik-a, of Germany. The Medlar.

**Metrosideros**, me-tros-id-*e*-ros; from Gr. *metra*, heart of a tree, and *sideros*, iron, alluding to the hardness of the wood. Greenhouse flowering shrubs.

FLORIBUNDA, flor-ib-*und*-a, free-flowering.
SPECIOSA, spes-e-*o*-sa, showy.

**Meum**, *me*-um, from Gr. *meion*, small, in allusion to the fineness of the foliage. Hardy aromatic herb.

ATHAMANTICUM, ath-am-*an*-tik-um, Athamanta-like, this genus being named after Mount Athamas.

**Meyenia**, mey-*en*-e-a; after M. Meyen. Greenhouse flowering shrubs.

ERECTA, e-*rek*-ta, erect habit.

**Michauxia**, me-*sho*-se-a; after André Michaux, a French botanist. Biennials.

CAMPANULOIDES, kam-pan-ul-*oy*-dez, campanula-like.

**Microglossa**, mi-kro-*glos*-sa; from Gr. *mikros*, small, and *glossa*, a tongue, presumably referring to the segments of the corolla. Shrub.

ALBESCENS, al-*bes*-senz, whitish.

**Microlepia**, mi-kro-*le*-pe-a; from Gr. *mikros*, small, and *lepis*, a scale, the appearance of the spore cases. Greenhouse ferns.

ANTHRISCIFOLIA, an-thris-ke-*fo*-le-a, leaved like Anthriscus (Beaked Parsley).
HIRTA CRISTATA, *her*-ta kris-*ta*-ta, hairy and crested.

**Micromeria**, mi-kro-*meer*-e-a; from Gr. *mikros*, small, and *meris*, a part, referring to the small flowers and leaves of these little shrubs. Half-hardy shrubby perennials.

CROATICA, kro-*at*-ik-a, of Croatia.
PIPERELLA, pi-per-*el*-la, small peppermint.
RUPESTRIS, roo-*pes*-tris, inhabits rocky places.

**Microsperma**, mi-kros-*per*-ma; possibly from Gr. *mikros*, small, and *sperma*, seed. Half-hardy annual.

BARTONIOIDES, bar-ton-e-*oy*-dez, Bartonia-like.

**Mikania**, mik-*an*-e-a; after Joseph Mikan, professor of botany at Prague. Greenhouse flowering climbers.

SCANDENS, *skan*-denz, climbing.

**Mila**, *mil*-a, anagram of Lima, in the neighbourhood of which the species is found. Greenhouse cacti.

CAESPITOSA, ses-pit-*o*-sa, tufted.

**Milla**, *mil*-la; after Julian Milla, a gardener to the royal house of Spain. Half-hardy bulbs.

BIFLORA, bif-*lo*-ra, two-flowered.
LAXIFLORA, laks-if-*lo*-ra, loose-flowered.
UNIFLORA, u-nif-*lo*-ra, one-flowered.

**Miltonia**, mil-*to*-ne-a; after Earl Fitzwilliam (Viscount Milton), patron of natural science and lover of orchids. Warm-house orchids.

BLEUANA, blu-*a*-na, after M. Alfred Bleu of Paris.
CLOWESII, *klowes*-ei, after the Rev. John Clowes.
PHALAENOPSIS, fal-e-*nop*-sis, moth-like.
ROEZLII, *roez*-lei, after M. Roezl.
SPECTABILIS, spek-*tab*-il-is, notable.
VEXILLARIA, veks-il-*la*-re-a, standard bearing—the conspicuous labellum.

**Mimosa**, mi-*mo*-sa; from Gr. *mimos*, imitator, referring to the sensitivity of the leaves to touch or injury. Greenhouse annuals. Sensitive plants.

PUDICA, *pud*-ik-a, humble, obedient.
SENSITIVA, sen-sit-*iv*-a, sensitive—to touch or injury.

**Mimulus**, *mim*-ul-us; from L. *mimus*, a mimic, the flowers being supposed to resemble a mask, or monkey's face, hence Monkey-flower, or from L. *mimo*, an ape. Herbaceous perennials; most kinds in cultivation are varieties of M. luteus.

CARDINALIS, kar-din-*a*-lis, cardinal, deep scarlet.
CUPREUS, *ku*-pre-us, copper-coloured.
GUTTATUS, gut-*ta*-tus, covered with small spots.
HARRISONII, har-ris-*on*-ei, after Harrison. The Harrison's Mask.
LUTEUS, *lu*-te-us, yellow. The Common Monkey-flower.
MOSCHATUS, mos-*ka*-tus, musk. The Musk Plant.
RINGENS, *rin*-gens, gaping—the flowers.

**Mina**, *my*-na, after Don Francisco Xavier Mina of Mexico. Greenhouse climbing annual.

LOBATA, lo-*ba*-ta, lobed—the leaves.

**Mirabilis,** mir-*ab*-il-is; from L. *mirabilis*, wonderful, or to be admired. Tender perennials.

JALAPA, jal-*a*-pa, old name. The Jalap, or Marvel of Peru.
LONGIFLORA, long-if-*lo*-ra, long-flowered.

**Miscanthus,** mis-*kan*-thus; from Gr. *miskos*, a stem, and *anthos*, a flower, referring to the tall flowering stems. Variegated leaved ornamental grasses.

SINENSIS UNIVITTATUS, sin-*en*-sis u-ne-vit-*ta*-tus, white-lined Chinese.
S. VARIEGATUS, var-e-eg-*a*-tus, variegated.
S. ZEBRINUS, ze-*bry*-nus, banded or cross-striped, like a zebra.

**Mitchella,** mit-*chel*-la, after Dr. John Mitchell, a botanist, of Virginia. Rock or woodland herb.

REPENS, re-penz, creeping.

**Mitella,** mit-*el*-la; the diminutive of L. *mitra*, a mitre (a little mitre), in allusion to the two-cleft seed pod. Small woodland plants.

DIPHYLLA, dif-*il*-la, two-leaved.

**Mitraria,** mit-*rah*-re-a; from L. *mitra*, referring to shape of bracts covering the calyx. Greenhouse shrub.

COCCINEA, kok-*sin*-e-a, scarlet.

**Molinia,** mol-*ee*-ne-a, after J. Molina, a writer on Chilean plants. Hardy grass.

CÆRULEA, se-*ru*-le-a, bluish green—the infloresence. Two variegated forms are grown—variegata and Bertini.

**Moltkia,** *molt*-ke-a; named after Count Moltke, a Danish nobleman. Shrubby perennial.

PETRÆA, pet-*re*-a, of rocks.

**Momordica,** mo-*mor*-dik-a; from L. *mordeo*, to bite, in allusion to the jagged seeds as though bitten. Tender annual fruiting climbers.

BALSAMINA, bawl-sam-*ee*-na, balsam scented.
CHARANTIA, kar-*an*-te-a, Charantia.
ELATERIUM, el-at-*e*-re-um, impelling—the seeds expelled from ripe fruits. The Squirting Cucumber.

**Monanthes,** mo-*nan*-thez; from Gr. *monos*, single, and *anthos*, a flower—one-flowered. Given in error as the type plant (polyphylla) was thought to be one-flowered; it is few-flowered. Greenhouse succulents.

LAXIFLORA, laks-if-*lo*-ra, loose-flowered.
MURALIS, mu-*ra*-lis, found on walls.
POLYPHYLLA, pol-if-*il*-la, many-leaved.

**Monarda,** mon-*ar*-da; after N. Monardez, a physician and botanist of Seville. Herbaceous perennials.

DIDYMA, *did*-im-a, double or twin, the flowers having stamens in pairs of different sizes.
FISTULOSA, fis-tu-*lo*-sa, hollow-stalked.

**Monilaria,** mo-nil-*ar*-ea; from L. *monila*, a necklace, the stems constricted into bead-like sections. Greenhouse succulents.

MONILIFORMIS, mo-nil-if-*or*-mis, necklace-like.

**Montbretia,** mont-*bre*-she-a; said to be named after M. Monbret, a French botanist. Hardy bulbous perennials. Most of the montbretias in common cultivation are of garden origin.

CROCOSMÆFLORA, kro-kos-me-*flor*-a, flowers like Crocosmia.
POTTSI, *potts*-e, after Potts.

**Monvillea,** mon-*vil*-le-a; after M. Monville, a student of cacti. Greenhouse cacti.

CAVENDISHI, kav-en-*dish*-e, after Cavendish.
SPEGAZZINI, speg-az-*zin*-e, after Spegazzin.

**Moræa,** mor-*ee*-a; after J. Moræus, a Swedish physician. Greenhouse bulbous plants.

BICOLOR, *bik*-ol-or, two-coloured.
ROBINSONIANA, rob-in-*so*-ne-a-na, after Robinson.
UNGUICULATA, un-gwik-ul-*a*-ta, narrow clawed—the lower end of the petal.

**Morina,** mor-*e*-na; named after L. Morin, a French plant collector and botanist. Herbaceous perennials.

COULTERIANA, kol-ter-e-*a*-na, after Coulter.
LONGIFOLIA, long-if-*o*-le-a, long-leaved.

**Morisia,** mor-*is*-e-a; after G. G. Moris, Italian botanist. Rock plant.

HYPOGÆA, hy-po-*je*-a, lit. "under the earth," the plant burying its seed capsules.
MONANTHOS, mo-*nan*-thos, one-flowered.

**Morus,** *mor*-us; ancient L. name. Tree cultivated for its edible fruit and leaves for silkworm forage.

ALBA, *al*-ba, white—the fruits.
NIGRA, *nig*-ra, black. The Mulberry.

**Moschosma,** mos-*koz*-ma; from Gr. *moschos,* musk, and *osme,* a smell—musk-like fragrance. Greenhouse flowering perennial.

RIPARIUM, re-*par*-e-um, frequenting river banks.

**Muehlenbeckia,** muhl-en-*bek*-e-a; after a Dr. G. Muehlenbeck. Slender trailing shrubs.

COMPLEXA, kom-*pleks*-a, interwoven, entangled—the branches. The Wire Vine.

**Mulgedium,** mul-*ge*-de-um; probably from L. *mulgeo,* to milk, the juice of the plants being white.

PLUMIERI, plu-me-*air*-e, after Plumier. The Blue Lettuce.

**Musa,** *mu*-za; origin doubtful, said to be in honour of Antonius Musa, a freedman of Emperor Augustus, whose physician he became; the Arabic and Egyptian name is Mauz and this is considered by some to be the basis of the Latin *Musa.* Tropical fruiting plants.

CAVENDISHII, kav-en-*dish*-ei, after Cavendish.

ENSETE, en-*se*-te, probably Abyssinian native name.

PARADISIACA, par-a-*dis*-e-ak-a, of Paradise. The Plantain.

SAPIENTUM, sap-e-*en*-tum, wise men's. The Banana.

**Muscari,** mus-*kar*-e; from Gr. *moschos,* musk, the flowers of some species having a musky odour. Hardy bulbs.

BOTRYOIDES, bot-re-*oy*-dez, like a bunch of grapes (Gr. *botrus*). The Grape Hyacinth.

COMOSUM, ko-*mo*-sum, hairy-tufted. The Tassel Hyacinth.

CONICUM, *ko*-nik-um, conical—the inflorescence.

MOSCHATUM, mos-ka-tum, musk-scented. The Musk Hyacinth.

RACEMOSUM, ras-e-*mo*-sum, flowers in racemes.

**Mussænda,** mus-*sen*-da; the Cingalese name for M. frondosa. Tropical flowering shrubs.

ERYTHROPHYLLA, er-ith-*rof*-il-la, red-leaved.

FRONDOSA, fron-*doz*-a, leafy.

MACROPHYLLA, mak-*rof*-il-la, large-leaved.

**Mutisia,** mew-*tis*-e-a; named after C. Mutis, a botanist of S. America. Greenhouse climbers.

CLEMATIS, *klem*-a-tis, clematis-like.

DECURRENS, de-*kur*-renz, running down—leaves clasping stems.

**Myosotidium,** my-o-so-*tid*-e-um; from *Myosotis* (see below) and Gr. *eidos,* an appearance, the flowers resembling those of Myosotis (forget-me-not). Herbaceous plant.

NOBILE, *no*-bil-e, noble.

**Myosotis,** my-o-*so*-tis; from Gr. *mus,* a mouse, and *otes,* an ear—the foliage resemblance. The plant to which the name was originally given by the ancient Greeks was not a forget-me-not. Hardy perennials or biennials.

ALPESTRIS, al-*pes*-tris, alpine.

AZORICA, az-*or*-ik-a, of the Azores.

DISSITIFLORA, dis-sit-if-*lo*-ra, the flowers far apart. The garden Forget-me-not.

PALUSTRIS, pal-*us*-tris, of marshes. The English Forget-me-not.

**Myrica,** mer-*ik*-a; Greek name (*myrike*) for a riverside shrub. Shrubs.

CERIFERA, ser-*if*-er-a, wax-bearing, the fruits yielding a candle-wax. The Candle-berry Myrtle.

GALE, gale, old name for Bog Myrtle (Sweet Gale), probably from *Gagel,* an Anglo-Saxon term for the same shrub.

**Myriophyllum,** mer-e-of-*il*-lum; from Gr. *myrios,* a myriad, and *phyllon,* a leaf, in reference to the many divisions in the leaves. Submerged aquatics. Water Milfoil and Feather-foil.

HETEROPHYLLUM, het-er-*of*-il-lum, varied-leaved, those emerged differing from those submerged.

PROSERPINACOIDES, pros-er-pin-ak-*oy*-deez, Proserpinaca-like. The Parrot's Feather.

SPICATUM, spe-*ka*-tum, spiked.

VERTICILLATUM, ver-tis-il-*la*-tum, whorled—the foliage.

**Myrrhis,** *mer*-ris; from Gr. *myrrha,* fragrant.

ODORATA, od-or-a-ta, fragrant.

**Myrsiphyllum,** mer-sif-*il*-lum; from Gr. *myrsine,* myrtle, and *phyllon,* a leaf, referring to the myrtle-like foliage.

ASPARAGOIDES, as-par-ag-*oy*-dez, Asparagus-like.

**Myrtus,** *mer*-tus; Latin name for a Myrtle (Gr. *murtos*), meaning obscure. Half-hardy shrubs.

COMMUNIS, kom-*mu*-nis, common.

LUMA, *leu*-ma, old name of Chilian origin.

UGNI, *ug*-ne (*u*-ne, of some), early generic name, Chilian.

**Nægelia,** nag-*el*-e-a; after Dr. Nageli, German botanist. Greenhouse tuberous herbaceous plants.

CINNABARINA, kin- (or sin-) nab-ar-*ee*-na, vermilion.

FULGIDA, *ful*-jid-a, shining.

MULTIFLORA, mul-tif-*lo*-ra, many flowered.

ZEBRINA, ze-*bry*-na, zebra-striped.

**Nananthus,** nan-*an*-thus; from Gr. *nanos,* dwarf, and *anthos,* a flower, alluding to the dwarfness of the plants. Greenhouse succulents.

POLE-EVANSII, pole-*ev*-anz-ei, after Pole Evans, a South African plant collector.

VITTATUS, vit-*ta*-tus, ribbon-striped.

**Nandina,** nan-*dy*-na; from *nandin,* the Japanese name for this half-hardy shrub.

DOMESTICA, do-*mes*-tik-a, domestic— from its variou uses in Japanese households.

**Narcissus,** nar-*sis*-sus; a classical L. name from the Gr. perhaps in allusion to narcotic qualities. Hardy bulbs; most garden kinds are hybrids.

BULBOCODIUM, bul-bo-*ko*-de-um, probably from Gr. *bolbos,* a bulb, and *kodion,* a little fleece, the covering of the bulb. The Hoop-petticoat Daffodil.

CYCLAMINEUS, sik-la-*min*-e-us, like a Cyclamen flower.

INCOMPARABILIS, in-kom-par-*a*-bil-is, incomparable.

JONQUILLA, jon-*kwil*-la, probably from L. *juncus,* a rush, the leaves being rush-like. The Jonquil.

JUNCIFOLIA, jun-sif- (or kif-) o-le-a, Juncus or rush-leaved.

MAJOR, *ma*-jor, larger.

MAXIMUS, *maks*-e-mus, largest.

MINOR, *mi*-nor, smaller.

ODORUS, od-*or*-us, sweet-scented.

POETICUS, po-*et*-ik-us, poet's. The Poets' Narcissus.

PSEUDO-NARCISSUS, *sued*-o-nar-*sis*-sus, the false narcissus. The English Daffodil.

TAZETTA, taz-*et*-ta, old name. Polyanthus Narcissus.

TRIANDRUS, tre-*an*-drus, having three stamens.

**Nasturtium,** nas-*tur*-she-um; from L. *nasi-tortium,* distortion of the nose, in reference to the pungency of some species. Aquatic salad herb.

**Nasturtium** (*continued*)

OFFICINALE, of-fis-in-*a*-le, of shops (herbal). The Water Cress.

The garden plant popularly called Nasturtium is an agnual Tropæolum (T. majus), which see.

**Neillia,** ne-*il*-le-a; named after Dr. Patrick Neill, a prominent Edinburgh botanist. Shrubs.

AMURENSIS, am-oor-*en*-sis, of Amurland.

OPUILFOLIA, op-ul-if-*o*-le-a, leaved like Viburnum Opulus, the Guelder-rose.

**Nelumbium,** ne-*lum*-be-um; from *nelumbo,* the Cingalese name for N. speciosum. Tropical aquatics.

LUTEUM, *lu*-te-um, yellow.

SPECIOSUM, spes-e-*o*-usm, showy.

**Nemesia,** nem-*e*-ze-a; old name used by Dioscorides for some sort of snapdragon. Half-hardy annuals used for bedding.

FLORIBUNDA, flor-ib-*un*-da, many-flowered.

STRUMOSUS, stroo-*mo*-sus, having tubercles.

VERSICOLOR, ver-*sik*-ol-or, changeable colour.

**Nemophila,** nem-*of*-il-a; from L. *nemos,* a grove or glade, and *phileo,* to love, the plants inhabiting such places. Hardy annuals.

INSIGNIS, in-*sig*-nis, showy. The Baby Blue-cyes.

MACULATA, mak-ul-*a*-ta, blotched.

**Nepenthes,** ne-*pen*-thez; from Gr. meaning without care, in allusion to the passage in the Odyssey where Helen so drugged the wine cup that its contents freed men from grief and care. The Pitcher Plant is sometimes grown in hothouses for curiosity. Tropical perennials.

CURTISII, *ker*-tis-ei, after Curtis.

DISTILLATORIA, dis-til-la-*tor*-e-a, distilling. The first introduced species.

NORTHIANA, north-e-*a*-na, after Miss North.

RAFFLESIANA, raf-les-e-*a*-na, after Sir Stamford Raffles.

VENTRICOSA, ven-trik-*o*-sa, swollen or bellied.

**Nepeta,** *nep*-e-ta (popularly nep-*e*-ta), possibly from Nepete, a city of Etruria. Hardy and half-hardy herbaceos plants.

GLECHOMA, glek-*o*-ma, Gr. *glechon,* a sort of thyme or pennyroyal; old

**Nepeta** (*continued*)
generic name for Ground Ivy. Variegated variety grown in gardens.
MUSSINII, moo-*se*-nei, after Mussin.

**Nephrodium**, nef-*ro*-de-um; from Gr. *nephros*, a kidney, the shape of the indusium covering the spore cases. Greenhouse ferns.
MOLLE, *mol*-le, soft to touch, downy.
RICHARDSII MULTIFIDUM, *rich*-ards-ei mul-*tif*-id-um, Richards' crested.
Many Lastreas are classed by some under Nephrodium.

**Nephrolepis**, nef-rol-*ep*-is; from Gr. *nephros*, a kidney, and *lepis*, a scale, referring to the shape of the indusium covering the spore cases. Greenhouse ferns.
BAUSEI, *baus*-e-e, after Bause.
CORDIFOLIA, kor-dif-*ol*-e-a, heart-shaped.
DAVALLIOIDES, dav-al-le-*oy*-dez, davallia-like.
EXALTATA, eks-al-*ta*-ta, elevated or tall.
RUFESCENS, ru-*fes*-enz, becoming red.
TODEOIDES, *to*-de-*oy*-dez, Todea-like.
Many finely divided varieties are known as Boston Ferns.

**Nerine**, ne-*ri*-ne; named after a princess of Grecian mythology. Greenhouse bulbs.
BOWDENII, bow-*den*-ei, after Bowden.
FLEXUOSA, fleks-u-*o*-sa, bent alternately or zigzag.
FOTHERGILLI, foth-er-*gil*-le, after Fothergill.
PUMILA, *pu*-mil-a, dwarf.
SARNIENSIS, sar-ne-*en*-sis, of Guernsey. The Guernsey Lily.

**Nerium**, *ne*-re-um; Gr. name of the oleander, possibly referring to the habitat of the plant. Greenhouse flowering shrubs.
OLEANDER, o-le-an-der, oleander.

**Nertera**, *ner*-ter-a; from Gr. *nerteros*, low down, the plant being very prostrate. Half-hardy herb.
DEPRESSA, de-*pres*-sa, pressed down. The Fruiting Duckweed.

**Nicandra**, nik-*an*-dra; after Nicander, a Greek physician, who wrote on plants. Hardy annual.
PHYSALOIDES, fy-sal-*oy*-dez, Physalislike—the seed vessels.

**Nicotiana**, nik-o-te-*a*-na (popularly ne-ko-she-*a*-na); said to be named after a French ambassador to Portugal named Jean Nicot, who is said to have first presented tobacco to the courts of Portugal and France. Annuals and perennials.
AFFINIS, af-*fin*-is, allied, presumably to N. alata; regarded by some as a synonym. The Night-scented Tobacco.
SANDERÆ, *san*-der-e, after Sander.
SUAVEOLENS, *swa*-ve-ol-enz, sweet smelling.
SYLVESTRIS, sil-*ves*-tris, sylvan or of woodland.
TABACUM, tab-*a*-kum, tobacco; old name for Virginian tobacco plant. The Tobacco.

**Nierembergia**, ne-er-em-*ber*-ge-a; after Juan E. Nieremberg, a Spanish Jesuit. Half-hardy perennials.
CÆRULEA, ser-*u*-el-a, blue.
FRUTESCENS, fru-*tes*-senz, shrubby.
HIPPOMANICA, hip-po-*man*-ik-a, land of horses or of meadows—the habitat.
RIVULARIS, riv-u-*lar*-is, of river-sides.

**Nigella**, ni-*jel*-la; from L. *nigellus*, the diminutive of *niger*, black, the very black seeds. Hardy annuals.
DAMASCENA, dam-as-*se*-na, Damascus, but word here probably means damask (same origin), the flowers and foliage of this plant suggesting a fine textile fabric. The Love in a Mist.
HISPANICA, his-*pan*-ik-a, of Spain.

**Niphobolus**, nif-*ob*-o-lus; from Gr. *niphos*, snow, and *bolos*, a large pill, referring to the white scaly covering on the underside of the frond and to the large round sori. Greenhouse fern.
LINGUA, *ling*-wa, a tongue—the shape of the fronds.

**Nolana**, no-*la*-na; from L. *nola*, a little bell, the form of the flowers. Half-hardy trailing annuals.
ATRIPLICIFOLIA, at-rip-lis-if-*o*-le-a, leaved like Atriplex.

**Nopalea**, no-*pal*-e-a; from Nopal, the Mexican name for some opuntias. Greenhouse cacti.
COCCINELLIFERA, kok-sin-el-*lif*-er-a, cochineal-bearing. The favourite food plant of the cochineal insect.

**Nopalxochia**, no-pal-*zok*-e-a; an Aztec name. Greenhouse cacti.
PHYLLANTHOIDES, fil-lanth-*oy*-dez, Phyllanthus-like.

**Nothochlæna,** noth-ok-*le*-na; from Gr. *nothos*, spurious, and *chlæna*, a cloak, referring to the imperfect indusium. Greenhouse ferns.

DEALBATA, de-al-*ba*-ta, whitewashed—the farina.
FLAVENS, *fla*-venz, yellow—the farina.
NIVEA, *niv*-e-a, white—the farina.
SINUATA, sin-u-*a*-ta, wavy.

**Nuphar,** *new*-far; from *naufar*, the Arabic name for this plant. Hardy aquatic.

LUTEUM, *lu*-te-um, yellow. The Yellow Water Lily.

**Nuttallia,** nut-*tal*-le-a; after Thomas Nuttall, a botanist of Cambridge. Deciduous flowering shrub.

CERASIFORMIS, ser-as-if-*or*-mis, cherry (Cerasus)-like.

**Nycterinia,** nik-ter-*in*-e-a; from Gr. *nykteros*, by night, in allusion to the night fragrance of the flowers.

CAPENSIS, ka-*pen*-sis, of Cape of Good Hope.
SELAGINOIDES, sel-ag- (or aj-) in-*oy*-dez, Selago-like.

**Nyctocereus,** *nik*-to-*se*-re-us; from Gr. *nyktos*, night, and *cereus*, the species are night blooming. Greenhouse cacti.

SERPENTINUS, ser-pen-*ti*-nus, creeping like a serpent—the coiling stems.

**Nymphæa,** *nim*-fe-a; from Gr. *Nymphe*, goddess of springs (water-nymph). The Water-Lilies. Most of those in cultivation are hybrids of garden origin.

ALBA, *al*-ba, white.
ODORATA, od-or-*a*-ta, sweet-scented.
PYGMÆA, *pig*-me-a, dwarf.
STELLATA, stel-*la*-ta, star-shaped—the flowers.
TUBEROSA, tu-ber-*o*-sa, tuberous.

**Ochna,** *ok*-na; from Gr. *ochne*, the wild pear—ochna leaves resemble those of the pear. Tropical flowering shrubs.

KIRKII, *kerk*-ei, after Kirk.
MULTIFLORA, mul-tif-*lo*-ra, many-flowered.

**Ocimum,** *o*-sim-um; an old Gr. name. Also spelled Ocymum. Annual aromatic herbs.

BASILICUM, ba-*sil*-ik-um, basil. The Sweet or Common Basil.
MINIMUM, *min*-e-mum, small, or least. The Bush Basil.

**Odontioda,** o-don-te-*o*-da; a name compounded from *Odonto*glossum and Cochl*ioda*, the species being hybrids between members of these two genera. Greenhouse orchids.

BRADSHAWIÆ, *brad*-shaw-e-e; after Mrs. Bradshaw.

**Odontoglossum,** o-don-to-*glos*-sum; from Gr. *odons*, a tooth, and *glossa*, a tongue, alluding to the shape of the lip. Greenhouse orchids.

ALEXANDRÆ, al-ex-*an*-dre, after Princess Alexandra.
BICTONENSE, bik-ton-*en*-se, of Bicton.
CERVANTESII, ser-van-*te*-sei, after Cervantes.
CITROSMUM, sit-*ros*-mum, lemon-scented.
CRISPUM, *kris*-pum, curled—the lip.
GRANDE, *gran*-de, magnificent.
HALLII, *hal*-lei, after Hall.
HARRYANUM, har-ry-*a*-num, after Sir Harry Veitch.
LUTEO-PURPUREUM, *lu*-te-o-pur-*pur*-e-um, yellow and purple.
PESCATOREI, pes-kat-*o*-re-ei, after Pescatore.
ROSSII, *ros*-sei, after Ross.
TRIUMPHANS, tre-*um*-fanz, triumphant.

**Œnanthe,** ee-*nan*-thee; from Gr. *oinos*, wine, and *anthos*, a flower—having a vinous odour. Aquatics.

FISTULOSA, fis-tul-*o*-sa, stem hollow between the joints or nodes. The Water Dropwort.
FLUVIATILIS, floo-ve-*a*-til-is, found in rivers.

**Œnothera,** ee-*noth*-e-ra (correctly en-oth-e-ra); from Gr. *oinos*, wine, and *thera*, pursuing or imbibing, the roots of one of the species (or an allied plant) being regarded by the Romans as an incentive to drinking. Herbaceous perennials, biennials, and annuals.

BIENNIS, bi-*en*-nis, biennial.
CÆSPITOSA, ses-pit-*o*-sa, tufted.
FRUTICOSA, frut-ik-*o*-sa, shrubby.
LAMARCKIANA, lam-*ark*-e-*a*-na, after Lamarck.
MISSOURIENSIS, mis-soor-e-*en*-sis, of Missouri.
PUMILA, *pu*-mil-a, dwarf.
SPECIOSA, spes-e-*o*-sa, showy.
TARAXACIFOLIA, tar-aks-a-sif-*o*-le-a, Taraxacum (Dandelion)-leaved.

**Olea,** *ol*-e-a; classical L. name. Greenhouse shrub.

EUROPÆA, u-ro-*pe*-a, European. The Olive Tree.

**Olearia**, ol-e-*a*-re-a; believed to refer to resemblance to Olea. Some species have olive-like leaves. Flowering shrubs.

GUNNIANA, gun-ne-*a*-na, after Gunn.

HAASTII, ha-*ast*-ei, after Haast.

MACRODONTA, mak-ro-*don*-ta, long-toothed—the leaves.

NUMMULARIÆFOLIA, num-mul-*ar*-e-e-*fo*-le-a, leaved like Lysimachia nummularia (Creeping Jenny).

STELLULATA, stel-lew-*la*-ta, a little star —the flowers.

**Omphalodes**, om-fa-*lo*-des; from Gr. *omphalos*, a navel, and *eidos*, like, the shape of the seeds. The Navelwort. Annuals and perennials.

CAPPADOCICA, kap-pa-*dos*-ik-a, of Cappadocia.

LINIFOLIA, lin-e-*fo*-le-a, Linum (flax)-leaved.

LUCILIÆ, lu-*sil*-e-e, after Mme. Lucile Boissier

NITIDA, *nit*-id-a, shining—the leaves.

VERNA, *ver*-na, spring—time of flowering.

**Oncidium**, on-*sid*-e-um; from Gr. for swelling or tubercle, alluding to the crest on lip. Warm-house epiphytal orchids.

BICALLOSUM, bik-al-*lo*-sum, two-warted.

CONCOLOR, *kon*-kol-or, one-coloured.

FLEXUOSUM, fleks-u-o-sum, zigzag—the flower stem.

FORBESII, *forbes*-ei, after Forbes.

KRAMERIANUM, kra-mer-e-*a*-num, after Kramer.

MACRANTHUM, mak-*ranth*-um, large-flowered.

ORNITHORHYNCHUM, or-ni-thor-*in*-kum, bird's bill or beak.

PAPILIO, pa-*pil*-e-o, butterfly-flowered.

ROGERSII, roj-*erz*-ei, after Rogers.

SPHACELATUM, sfak-el-*a*-tum, scorched.

SPLENDIDUM, splen-*did*-um, splendid.

TIGRINUM, tig-*re*-num, tiger-striped.

VARICOSUM, var-ik-*o*-sum, dilated veins.

**Onoclea**, on-*ok*-le-a; from Gr. *onokleia*, a plant with leaves rolled up into the semblance of berries, referring to the capsule-like fructification.

GERMANICA, jer-*man*-ik-a, of Germany.

SENSIBILIS, sen-*sib*-il-is, sensitive. The Sensitive Fern.

**Ononis**, on-*o*-nis; ancient Gr. name. Rock and border perennials.

FRUTICOSA, frut-ik-*o*-sa, shrubby.

HIRSINA, hir-*si*-na, with goat smell.

MINUTISSIMA, min-*u*-tess-ima, very small.

**Ononis** (*continued*)

NATRIX, *nat*-riks, goat-root.

ROTUNDIFOLIA, ro-tun-dif-*o*-le-a, round-leaved.

**Onopordum**, on-op-*or*-dum; old Gr. name, possibly from Gr. *onos*, an ass, and *perdo*, to consume, referring to asses eating thistle foliage. Ornamental thistles.

ACANTHIUM, ak-*anth*-e-um, Acanthus-like. The Cotton Thistle.

ARABICUM, ar-*ab*-ik-um, of Arabia.

ILLYRICUM, il-*ler*-ik-um, of Illyria.

**Onosma**, on-*oz*-ma; from Gr. *onos*, an ass, and *osme*, a smell, reference obscure. Rock plants.

ALBO-ROSEUM, *al*-bo-ro-*ze*-um, white and rosy.

TAURICUM, *taw*-rik-um, of Taurus. The Golden Drop.

**Onychium**, on-*ik*-e-um; from Gr. *onychion*, a little nail, referring to the shape of the fertile segments. Greenhouse ferns.

JAPONICUM, jap-*on*-ik-um, of Japan.

**Ophiopogon**, of-e-op-*o*-gon, from Gr. *ophis*, a serpent, and *pogon*, a beard. Greenhouse perennials; variegated varieties have ornamental foliage.

JABURAN, jab-*u*-ran, oriental name.

JAPONICUS, jap-*on*-ik-us, of Japan.

SPICATIUS, spe-*ka*-tus; bearing spikes.

**Ophrys**, *of*-ris; from Gr. *ophrys*, eyebrows, the allusion being to the fringe of the inner sepals. Terrestrial orchids.

APIFERA, ap-*if*-er-a, bee-bearing. The Bee Orchis.

ARANIFERA, ar-a-*nif*-er-a, spider-bearing.

MUSCIFERA, mus-*sif*-er-a, fly-bearing. The Fly Orchis.

TENTHREDINIFERA, ten-thre-din-*if*-er-a, sawfly-bearing.

**Oplismenus**, op-*lis*-me-nus; from Gr. *hoplismenos*, awned, referring to the awned glumes in the inflorescence. Greenhouse ornamental foliage grass — variegated variety.

BURMANNI, bur-*man*-ne, after Burmann.

Variety variegatus usually grown.

**Opuntia**, o-*pun*-te-a; from Opuntus (or Opus), a town in Greece; it is also said the word is a Latin one not applicable to the plants now so named. Greenhouse cacti.

ENGELMANNII, en-gel-*man*-nei, after Engelmann.

**Opuntia** (*continued*)

FICUS-INDICA, *fi*-kus-*in*-dik-a, Indian fig. The Indian Fig.

LEUCOTRICHA, lu-ko-*trik*-a, white-haired.

MACRORHIZA, mak-rorh-*e*-za, large rooted.

MESACANTHA, mes-ak-*anth*-a, middle spined.

MICRODASYS, mi-*krod*-as-is, small and thick.

MONACANTHA, mon-ak-*anth*-a, one-spined.

MISSOURIENSIS, mis-soor-e-*en*-sis, of Missouri.

PAPYRACANTHA, pap-y-rak-*anth*-a, papery spined.

RAFINESQUEI, ra-fin-*es*-kwi, after Rafinesque.

TUNA, *tu*-na, Tuna, native name.

VULGARIS, vul-*gar*-is, common.

**Orchis**, *or*-kis; ancient Gr. name, possibly referring to the two oblong tubers at the root of many species. Hardy terrestrial orchids.

FOLIOSA, fo-le-*o*-sa, profusely leaved, notably the flower-stalks.

HIRCINA, her-*se*-na, pertaining to goats—the smell.

LAXIFLORA, laks-if-*lor*-a, loose-flowered.

MACULATA, mak-ul-*a*-ta, spotted.

MASCULA, *mas*-ku-la, male, probably referring to its earliness and vigour.

PURPUREA, pur-*pur*-e-a, purple.

PYRAMIDALIS, pir-am-*id*-al-is, pyramid or cone-shaped—the flower spike.

SAMBUCINA, sam-*bu*-kin-a, elder (Sambucus)-scented.

**Oreocereus**, *or*-e-o-*se-re*-us; from Gr. *oros*, a mountain, and *cereus* (*q.v.*). Greenhouse cacti.

CELSIANUS, kel-se-*a*-nus, Celsian.

TROLLI, *trol*-le, after Troll. The Old Man of the Andes cactus.

**Origanum**, or-*ig*-a-num; from Gr. *oros*, a mountain, and *ganos*, beauty, the usual habitat and attractiveness of the plants. Herbaceous and sub-shrubby perennials.

DICTAMNUS, dik-*tam*-nus, old generic name. The Dittany of Crete.

MARJORANA, mar-jor-*a*-na, marjoram, old name. The Sweet Marjoram.

ONITES, on-*e*-tez, onites. The Pot Marjoram.

VULGARE, vul-*gar*-e, common. The Common Marjoram.

**Ornithogalum**, or-nith-*og*-a-lum; from Gr. *ornis*, a bird, and *gala*, milk; "bird's milk" was said to be a current expression among the ancient Greeks

**Ornithogalum** (*continued*)

for some wonderful thing. Bulbous plants.

ARABICUM, ar-*ab*-ik-um, of Arabia.

LACTEUM, *lak*-te-um, milk-white.

LONGIBRACTEATUM, long-e-brak-te-*a*-tum, with long bracts.

NARBONENSE, nar-bon-*en*-se, of Narbonne.

NUTANS, *nu*-tans, nodding.

PYRAMIDALE, pir-am-*id*-al-e, pyramidal—the flower spike.

PYRENAICUM, pir-en-*a*-ik-um, Pyrenean.

THYRSOIDES, ther-*soy*-dez, flowers in a thyrse-like spike.

UMBELLATUM, um-bel-*la*-tum, umbelled. The Star of Bethlehem.

**Ornus**, *or*-nus; old Gr. and L. name for Mountain Ash. Trees.

VULGARIS, vul-*gar*-is, common.

**Orobus**, *or*-o-bus; old Gr. name, said to be derived from Gr. *oro*, to excite, and *bous*, an ox, the vetches being tempting fodder. Herbaceous perennials and rock plants.

VERNUS, *ver*-nus, vernal or spring—time of flowering.

**Orontium**, or-*on*-te-um; old Greek name for a plant that grew on the banks of the river Orontes. Aquatic perennial.

AQUATICUM, a-*kwat*-ik-um, aquatic.

**Osmanthus**, oz-*man*-thus; from Gr. *osme*, perfume, and *anthos*, a flower, the blossoms being very fragrant. Shrubs.

AQUIFOLIUM, ak-we-*fo*-le-um, holly-leaved.

DELAVAYI, del-a-*vay*-i, after Delavay.

FRAGRANS, *fra*-granz, fragrant.

**Osmunda**, oz-*mun*-da; various derivations, all legendary, have been offered in explanation of this name, the most probable being a Saxon word signifying strength, Osmunda regalis being tall and robust. Another explanation is from Osmunder, one of the names of the Scandinavian Thor. Hardy ferns.

CINNAMOMEA, sin-na-*mo*-me-a, cinnamon, the colour of the fructification.

CLAYTONIANA, kla-to-ne-*a*-na, after Clayton, an American botanist.

REGALIS, re-*ga*-lis, royal, stately. The Royal Fern.

**Ostrowskia**, os-*trow*-ske-a; named after Ostrowsky, a Russian botanist. Hardy perennial.

MAGNIFICA, mag-*nif*-ik-a, magnificent.

**Ostrya**, *os*-tre-a; Possibly from Gr. *ostryos*, a scale, in allusion to the scaly catkins. Deciduous trees.

    CARPINIFOLIA, kar-pi-nif-*ol*-e-a, Carpinus (hornbeam)-leaved.

**Othonna**, oth-*on*-na; ancient Gr. name, possibly from Gr. *othone*, linen, the leaves of many of these ragworts having a soft downy covering. Greenhouse herbaceous perennials and shrubs.

    ARBORESCENS, ar-bor-*es*-senz, tree-like.
    CORONOPIFOLIA, kor-on-op-e-*fo*-le-a, Coronopus-leaved.
    CRASSIFOLIA, kras-sif-*o*-le-a, thick-leaved.
    DENTICULATA, den-tik-u-*la*-ta, toothed —the leaves.
    DIGITATA, dij-e-*ta*-ta, fingered—the leaves.
    FILICAULIS, fil-e-*kaw*-lis, thread-stemmed.
    HETEROPHYLLA, het-er-of-*il*-la, variously leaved.
    PERFOLIATA, per-fo-le-*a*-ta, the leaf-stem pierced.

**Othonnopsis**, oth-on-*nop*-sis; from Othonna (*q.v.*) and *opsis*, a resemblance. Half-hardy sub-shrub.

    CHEIRIFOLIA, ky-rif-*o*-le-a, wallflower-leaved.

**Ourisia**, owr-*is*-e-a; after Governor Ouris of the Falkland Islands. Hardy rock plants.

    COCCINEA, kok-*sin*-e-a, scarlet.

**Oxalis**, *ox*-a-lis; from Gr. *oxis*, acid, alluding to the acidity of the leaves of many species. Rock garden and woodland perennials.

    ACETOSELLA, as-et-o-*sel*-la, old generic name, from L. *acetum*, sour, the leaves. The Wood Sorrel.
    ADENOPHYLLA, ad-en-*of*-il-la, with glandular leaves.
    BOWIEI, *bo*-e-i, after Bowie.
    CERNUA, *ser*-nu-a, drooping, the flowers.
    CORNICULATA RUBRA, kor-nik-ul-*a*-ta *roo*-bra, red-leaved and horned.
    DEPPEI, *dep*-pe-i, after Deppe, a botanist.
    ENNEAPHYLLA, en-ne-*af*-il-la, nine-leaved, *i.e.*, nine divisions to each leaf.
    FLORIBUNDA, flor-ib-*un*-da, flowering abundantly.
    LASIANDRA, las-e-*an*-dra, woolly stamened.
    LOBATA, lo-*ba*-ta, lobed.
    OREGANA, or-e-*ga*-na, from Oregon.
    TETRAPHYLLA, tet-raf-*il*-la, four-leaved (leafleted).
    VALDIVIANA, val-div-e-*a*-na, Valdivian.

**Oxera**, ox-*e*-ra; from Gr. *oxeros*, sour, referring to the taste. Warm-house flowering climber.

    PULCHELLA, pul-*kel*-la, pretty.

**Oxycoccos**, oks-e-*kok*-kos; from Gr. *oxys*, sharp or bitter, and *kokkos*, a berry, in reference to the sourness of the fruits.

    MACROCARPUS, mak-ro-*kar*-pus, large-fruited.
    PALUSTRIS, pal-*us*-tris, marsh-loving. The Cranberry.

**Oxypetalum**, ox-e-*pet*-a-lum; from Gr. *oxys*, sharp, and *petalon*, a petal—the petals sharp pointed. Warm-house flowering climbers.

    CÆRULEUM, ser- (or ker-) *u*-le-um, sky-blue.

**Oxytropis**, ox-e-*tro*-pis; from Gr. *oxys*, sharp, and *tropis*, a keel, the keel petals ending in a sharp point. Hardy herbaceous perennials.

    MONTANA, mon-*ta*-na, of mountains.
    PYRENAICA, pir-en-*a*-ik-a, Pyrenean.
    URALENSIS, u-ra-*len*-sis, from the Ural Mountains.

**Ozothamnus**, o-zo-*tham*-nus; from Gr. *ozein*, to smell, and *thamnos*, a shrub, alluding to the odour. Flowering shrub.

    ROSMARINIFOLIUS, roz-mar-in-if-*ol*-e-us, rosemary (Rosmarinus)-leaved.

**Pachycereus**, *pak*-e-*se*-re-us; from Gr. *pachys*, thick, and *cereus* (*q.v.*). Cacti.

    PRINGLEI, *prin*-gl-e, after Pringle.

**Pachysandra**, pak-is-*an*-dra; from Gr. *pakys*, thick, and *aner*, a man, alluding to the unusually thick stamens. Woodland undershrubs.

    PROCUMBENS, pro-*kum*-benz, procumbent, *i.e.*, trailing without rooting.
    TERMINALIS, ter-min-*a*-lis, flowers terminal.

**Pæonia**, pe-*o*-ne-a (English rendering, Peony, *pe*-O-ne); named after Pæon, a physician of ancient Greece who first used the plant medicinally. Most of the garden peonies are varieties or hybrids.

    ALBIFLORA, al-bif-*lo*-ra, white-flowered.
    CORALLINA, kor-al-*le*-na, the colour of coral.
    LUTEA, *lu*-te-a, yellow.
    MOUTAN, *moo*-tan, Japanese name, derived from *Meu-tang*, the King of Flowers in Chinese mythology. The Tree Peony.

**Pæonia** (*continued*)

OFFICINALIS, of-fis-in-*a*-lis, of the shop (herbal). Var. rubra is the Double Red Peony Rose of cottage gardens.

**Palava** (or **Palaua**), pal-*a*-va; after Anton Palau y Verdera, Spanish professor of botany. Half-hardy annual.

DISSECTA, dis-*sek*-ta, cut—the leaves.

FLEXUOSA, flex-u-*o*-sa, crooked or zigzag —the stems.

**Paliurus**, pal-e-*u*-rus; the Gr. name used by Theophrastus. Flowering shrubs.

ACULEATUS, ak-u-le-*a*-tus, thorny. The Christ's Thorn.

**Panax**, *pan*-aks; from Gr. *pan*, all, and *akos*, a remedy, in allusion to the drug obtained from P. quinquefolium (Ginseng), to which miraculous virtues have been ascribed. Tropical ornamental-leaved shrubs.

BALFOURII, *bal*-four-ei, after Balfour.

FILICIFOLIUM, fil-is-if-*o*-le-um, ferny-leaved.

GUILFOYLEI, gwil-*foy*-le-i, after Guil-foyle.

VICTORIÆ, vik-*tor*-e-e, after Queen Victoria.

**Pancratium**, pan-*kra*-te-um; from Gr. *pan*, all, and *kratos*, potent, alluding to the use of these plants in ancient medicine. Greenhouse bulbous plants.

FRAGRANS, *fra*-granz, fragrant.

ILLYRICUM, il-*lir*-ik-um, of Illyria, on the Adriatic.

MARITIMUM, mar-*it*-im-um, maritime.

**Pandanus**, *pan*-dan-us; from Pandang, the Malay name. Tropical foliage plants. The Screw Pine.

BAPTISTII, bap-*tis*-tei, after Baptist.

SANDERI, *san*-der-i, after Sander.

VARIEGATUS, var-e-eg-*a*-tus, variegated.

VEITCHII, *veech*-ei, after Harry Veitch.

**Panicum**, *pa*-nik-um; the Latin name for a kind of millet used for bread-making. Ornamental grasses.

CAPILLARE, kap-il-*lar*-e, hair-like, *i.e.*, slender.

PLICATUM NIVEO-VITTATUM, plik-*a*-tum niv-e-o-vit-*ta*-tum, snowy striped and plaited.

VARIEGATUM, var-e-eg-*a*-tum, variegated.

**Papaver**, pap-*a*-ver; Latin name for a Poppy, of doubtful origin. Annuals and perennials.

ALPINUM, al-*pine*-um, alpine.

GLAUCUM, *glaw*-kum, glaucous or blue-green—the foliage. The Tulip Poppy.

**Papaver** (*continued*)

NUDICAULE, nu-dik-*aw*-le, naked stemmed. The Iceland Poppy.

ORIENTALE, or-e-en-*ta*-le, eastern. The Oriental Poppy.

PAVONINUM, pa-vo-*ne*-num, a peacock. The Peacock Poppy.

RHŒAS, *re*-as, old generic name, possibly from Gr. *rhoia*, a pomegranate, which the flower and fruit of the field poppy were supposed to resemble. The Corn Poppy. Shirley strain developed from this type.

RUPIFRAGUM, roo-pe-*frag*-um, rock-breaking.

SOMNIFERUM, som-*nif*-er-um, causing sleep. The Opium Poppy. Carnation and other strains developed from this type.

UMBROSUM, um-*bro*-zum, growing in shady places.

**Papyrus**, pa-*py*-rus; origin of the name is obscure. Greenhouse waterside perennial.

ANTIQUORUM, an-te-*kor*-um, ancient.

**Paradisia**, par-a-*dis*-ea, commemorates Giovanni Paradisi, an Italian. Hardy herbaceous plants.

LILIASTRUM, lil-e-*as*-trum, star lily. St. Bruno's Lily.

**Paris**, *par*-is; from L. *par*, equal, referring to the regularity of the parts. Hardy perennial.

QUADRIFOLIA, kwad-rif-*o*-le-a, four-leaved.

**Parnassia**, par-*nas*-se-a; named after Mount Parnassus, a sacred mountain of the ancient Greeks, from which these waterside plants were fabulously supposed to have sprung.

PALUSTRIS, pal-*us*-tris, of marshes. The Grass of Parnassus.

**Parochetus**, par-*ok*-e-tus; from Gr. *para*, near, and *achetos*, a brook, the plant delighting in watersides. Carpeting herb.

COMMUNIS, kom-*mu*-nis, communal, *i.e.*, growing in society.

**Paronychia**, par-o-*nik*-e-a; from Gr. *paronuchia*, a whitlow, which these plants were supposed to cure, hence English names Nailwort and Whitlow-wort.

ARGENTEA, ar-*jen*-te-a, silvery.

CAPITATA, kap-it-*a*-ta, flowers clustered in heads.

SERPYLLIFOLIA, ser-pil-lif-*o*-le-a, thyme-leaved.

**Parrotia,** par-*ro*-te-a; named after F. W. Parrot, German naturalist and traveller. Trees.

PERSICA, *per*-sik-a, of Persia.

**Passiflora,** pas-se-*flor*-a; from L. *passus*, suffering, and *flos*, a flower, lit. the Flower of the Passion (Passion-Flower), the early Spanish R.C. priests in South America finding in the plants features they regarded as symbols of the Crucifixion. Thus the five stamens were the five wounds; the three stigmas, the three nails; the style of the pistil, the flogging column; the corona, the crown of thorns or the halo of glory; the digitate or fingered leaves, the hands of the multitude; the coiled tendrils, the flogging cords; the five sepals and five petals, the ten disciples (Peter and Judas being omitted in the count). Greenhouse climbers.

ALATA, a*l*-*a*-ta, winged, the stalks.
CÆRULEA, ser- (or ker-) *u*-le-a, sky-blue.
COCCINEA, kok-*sin*-e-a, scarlet.
EDULIS, ed-*u*-lis, eatable or edible.
INCARNATA, in-kar-*na*-ta, flesh-coloured.
PRINCEPS, *prin*-seps, princely.
RACEMOSA, ras-e-*mo*-sa, racemose-flowered.
QUADRANGULARIS, kwod-ran-gul-*a*-ris, square-stalked.
VITIFOLIA, vi-tif-*o*-le-a, vine-leaved.

**Patrinia,** pat-*rin*-e-a; commemorating M. Patrin, French botanist. Herbaceous perennials and biennials.

PALMATA, pal-*ma*-ta, the leaves palmate, or hand-shaped.
SCABIOSIÆFOLIA, ska-be-*o*-se-*fo*-le-a, scabious-leaved.
VILLOSA, vil-*lo*-sa, shaggy.

**Paullinia,** paul-*lin*-e-a; after a Danish botanist, Paulli. Greenhouse flowering plant.

THALICTRIFOLIA, thal-ik-trif-*ol*-e-a, Thalictrum-leaved.

**Paulownia,** paw-*lo*-ne-a; named after Anna Paulowna, a princess of the Netherlands. Flowering and foliage tree.

IMPERIALIS, im-peer-e-*a*-lis, imperial.

**Pavia,** *pa*-ve-a; named after P. Paw, a Dutch botanist. Flowering trees.

CALIFORNICA, kal-if-*or*-nik-a, of California.
CARNEA, *kar*-ne-a, flesh-coloured. Buckeye.

**Pavonia,** pav-*o*-ne-a; after Don José Pavon, M.D., of Madrid, a traveller. Tropical flowering plants.

COCCINEA, kok-*sin*-e-a, scarlet.
MULTIFLORA, mul-tif-*lo*-ra, many-flowered.

**Pedicularis,** ped-ik-u-*lar*-is; from L. *pediculus*, a louse, the plant (Lousewort) being supposed to cause lice to appear on sheep which browsed upon it. Rock plants.

DOLICHORHIZA, dol-ik-orh-*e*-za, long-rooted.
FLAMMEA, *flam*-me-a, flame-coloured.
MEGALANTHA, meg-al-*anth*-a, large-flowered.
SCEPTRUM-CAROLINUM, *sep*-trum-kar-o-*lin*-um, Charles's Sceptre. (King Charles XII of Sweden.)

**Pediocactus,** *ped*-e-o-*kak*-tus; from Gr. *pedios*, a plain, and *cactus*, referring to the habitat—the Great Plain of Colorado. Hardy or greenhouse cacti.

SIMPSONII, *sim*-son-ei, after Simpson.

**Pelargonium,** pel-ar-*go*-ne-um; from Gr. *pelargos*, a stork, the ripe seed-head being supposed to resemble the head and beak of that bird, hence Storksbill. Most kinds commonly grown are of garden origin and are popularly spoken of as "Geraniums."

CAPITATUM, kap-it-*a*-tum, round-headed.
CITRIODORUM, sit-re-od-*or*-um, citron-scented.
CRISPUM, *krisp*-um, curled-leaved.
ECHINATUM, ek-in-*a*-tum, prickly.
GRANDIFLORUM, gran-dif-*lo*-rum, large-flowered. The Regal Pelargonium.
GRAVEOLENS, *grav*-e-ol-enz, strong smelling.
INQUINANS, *in*-kwe-nanz, dyed or stained.
PELTATUM, pel-*ta*-tum, shield-like. The Ivy-leaved Pelargonium.
RADULA, rad-*u*-la, rasp-leaved.
TOMENTOSUM, to-men-*to*-sum, downy-leaved.
ZONALE, zo-*na*-le, zoned or banded. The Zonal Pelargonium.

**Pelecyphora,** pel-ek-if-*or*-a; from Gr. *pelekyphorus*, hatchet-bearing, the shape of the tubercles suggesting hatchets. Greenhouse cacti.

ASELLIFORMIS, as-*el*-lif-*or*-mis, woodlouse-like—the tubercles.

**Pellæa,** pel-*le*-a; from Gr. *pellos*, dark coloured, referring to the black stems or stipes of the fronds. Greenhouse ferns.

**Pellæa** (*continued*)

ATROPURPUREA, at-ro-pur-*pur*-e-a, dark purple.

CALOMELANOS, kal-om-*el*-an-o·, beautiful black—stems or stipes.

FLEXUOSA, flex-u-*o*-sa, zig-zag—the fronds.

GERANIÆFOLIA, jer-a-ne-e-*fol*-e-a, geranium-leaved.

HASTATA, has-*ta*-ta, armed as with spears or halberds—the shape of pinnæ.

ROTUNDIFOLIA, ro-tun-dif-*o*-le-a, round leaved—the pinnæ.

TERNIFOLIA, ter-nif-*ol*-e-a, leaves in threes —the pinnæ.

**Peltandra**, pel-*tan*-dra; from Gr. *pelte*, a little shield, and *aner*, a man, the united stamens having the form of a shield. Aquatics.

VIRGINICA, ver-*jin*-ik-a, Virginian.

**Pennisetum**, pen-nis-*e*-tum; from L. *penna*, a feather, and *seta*, a bristle, the hairs attached to the flower plumes being feathered in some species. Ornamental grasses.

LONGISTYLUM, long-is-*ti*-lum, long-styled.

SETOSUM, se-*to*-sum, bristly.

**Penstemon**, a synonym of Pentstemon, which see.

**Pentas**, *pen*-tas; from Gr. *pente*, five, the parts of the flower being in fives. Tropical evergreen flowering shrubs.

CARNEA, *kar*-ne-a, flesh-coloured.

**Pentstemon**, pent-*ste*-mon, from Gr. *pente*, five, and *stemon*, a stamen, alluding to the five (four fertile and one rudimentary) stamens. Hardy perennials; many hybrids.

BARBATUS, bar-*ba*-tus, bearded.

BRIDGESII, brid-*je*-zei, after Bridges, an American collector.

CAMPANULATUS, kam-pan-u-*la*-tus, bell-flowered.

COBÆA, *ko*-be-a, cobæa-like—the flowers.

CONFERTUS, kon-*fer*-tus, closely crowded —the flowers.

CORDIFOLIUS, kor-dif-*o*-le-us, heart-shaped leaves.

DIFFUSUS, dif-*few*-sus, loosely spread in habit.

GLABER, *gla*-ber, smooth.

GLAUCUS, *glaw*-kus, blue-grey.

HARTWEGII, hart-*ve*-gei, after Hartweg.

HETEROPHYLLUS, het-er-of-*il*-lus, variously-leaved.

HUMILIS, *hu*-mil-is, lowly.

ISOPHYLLUS, i-so-*fil*-lus, equal-leaved.

**Pentstemon** (*continued*)

MENZIESII, men-*ze*-zei (Scotch *ming*-is), after Menzies.

RŒZLII, rez-lei, after Rœzl, a botanist and collector.

RUPICOLA, roo-*pik*-o-la, of rocks.

SCOULERI, *skool*-er-i, after Scouler.

SECUNDIFLORUS, se-kun-dif-*lor*-us, a one-sided flower spike.

TORREYI, *tor*-re-i, after Torrey, an American botanist.

**Peperomia**, pep-er-*o*-me-a; from Gr. *piper*, pepper, and *omoios*, similar— flowers and foliage similar to those of the pepper plant. Tropical ornamental leaved plants.

ARGYREIA, ar-ger-*i*-a, silver-striped.

MARMORATA, mar-mor-*a*-ta, marbled.

NUMMULARIÆFOLIA, num-mul-*ar*-e-e-*fo*-le-a, moneywort-leaved.

SANDERSII, san-derz-ei, after Sanders.

**Pereskia**, per-*esk*-e-a; from Nicolas F. Peiresc, a French patron of botany. Greenhouse cacti.

ACULEATA, ak-u-le-*a*-ta, prickly.

BLEO, *ble*-o, Bleo—the native name.

**Perezia**, pe-*re*-se-a; after Lazarus Perez, apothecary of Toledo. Half-hardy biennial.

MULTIFLORA, mul-tif-*lo*-ra, many flowered.

**Perilla**, per-*il*-la; the native Indian name. Half-hardy bedding annual.

NANKINENSIS, nan-kin-*en*-sis, from Nankin.

**Periploca**, per-*ip*-lo-ka; from Gr. *periploke*, an intertwining, the habit of the plant. Hardy twiner.

GRÆCA, *gre*-ka, of Greece.

**Peristeria**, per-is-*teer*-e-a; from Gr. *peristera*, a dove, in allusion to the form of the column. Warm-house orchids.

ELATA, e-*la*-ta, high, *i.e.*, tall growing.

**Pernettya**, per-*net*-e-a; named after Don Pernetty, author of a book on the Falkland Islands. Berry-bearing.

MUCRONATA, muk-ron-*a*-ta, leaves terminating in a point or mucro.

**Perowskia**, per-*ow*-ske-a; after M. Perowsky, once Governor of Russian province of Ouenberg. Sub-shrubby perennial.

ATRIPLICIFOLIA, at-rip-lis-e-*fo*-le-a, Atriplex-leaved.

**Persea,** *per*-se-a; ancient name of an oriental tree. Tropical fruiting shrub.

GRATISSIMA, gra-*tis*-sim-a, most grateful. The Avacado Pear.

**Persica,** *per*-sik-a; from *persicum*, the Latin name for peach, literally Persia, whence the tree was supposed to have come.

DAVIDIANA, da-vid-e-*a*-na, after Abbé David.

VULGARIS, vul-*gar*-is, common. The Peach.

V. LÆVIS, *la*-vis, smooth. The Nectarine.

**Petasites,** pet-a-*se*-tes; from Gr. *petasos*, a broad-brimmed hat, or sunshade, in reference to the large leaves of some species. Herbaceous perennials.

FRAGRANS, *fra*-granz; fragrant. The Winter Heliotrope.

**Petrea,** pet-*re*-a; after Robert James, Lord Petre, a patron of botany. Greenhouse flowering climber.

VOLUBILIS, vol-*u*-bil-is, twisting round— the twining stems.

**Petunia,** pe-*tu*-ne-a; from *petun*, a native Brazilian name for tobacco, petunias being allied to the Tobacco Plant. Greenhouse and bedding plants.

NYCTAGINIFLORA, nik-ta-gin-if-*lo*-ra, Nyctaginia-flowered.

VIOLACEA, vi-ol-*a*-se-a, violet-coloured. Garden petunias are hybrids of these two species.

**Peucedanum,** pu-*sed*-a-num; the Greek name used by Hippocrates.

GRAVEOLENS, grav-e-*ol*-enz, strong smelling.

SATIVUM, *sat*-iv-um, cultivated. The Parsnip.

**Phacelia,** fa-*se*-le-a; from Gr. *phakelos*, a bundle, alluding to the disposition of the flowers. Annuals and perennials.

CAMPANULARIA, kam-pan-u-*lar*-e-a, bell-shaped, the flowers.

PARRYI, *par*-re-e, after Parry, botanist, U.S.A.

TANACETIFOLIA, tan-a-set-if-*o*-le-a, tansy (Tanacetum)-leaved.

VISCIDA, *vis*-kid-a, viscid or sticky—the stems and leaves.

WHITLAVIA, whit-*la*-ve-a, old generic name, which see.

**Phaius,** *fa*-us or *fa*-jus; from Gr. for dark and swarthy, from colour of flowers of original species. Warm-house terrestrial orchids.

**Phaius** (*continued*)

BLUMEI, *blu*-me-e, after Blume.

COOKSONII, kook-*so*-nei, after Cookson.

GRANDIFOLUS, gran-dif-*ol*-e-us, large-leaved.

TUBERCULOSIS, tu-ber-kul-*o*-sis, tubercled.

**Phalænopsis,** fal-a-*nop*-sis; from Gr. *phalaina*, a moth, and *opsis*, resemblance, the flowers resembling moths or butterflies. Warm-house orchids.

AMABILIS, am-a-bil-is, lovely.

GRANDIFLORA, gran-dif-*lo*-ra, large-flowered.

LOWII, *low*-ei, after Low.

SANDERIANA, san-der-e-*an*-a, after Sander, nurseryman.

SCHILLERIANA, shil-ler-e-*an*-a, after Schiller.

**Phalaris,** *fal*-ar-is; ancient Gr. name, possibly from *phalaros*, shining, referring to the polished seeds. Ornamental grasses.

ARUNDINACEA, ar-un-din-*a*-se-a, reed (Arundo)-like.

CANARIENSIS, kan-ar-e-*en*-sis, providing canary seed.

**Pharbitis,** far-*bi*-tis; from Gr. *pharbe*, colour, in allusion to the richly-tinted flowers; a disused synonym of Ipomea.

**Phaseolus,** *fas*-e-o-lus; ancient name. A wide range of annuals and perennials including climbing and dwarf beans.

MULTIFLORUS, mul-tif-*lo*-rus, many flowered. The Scarlet Runner Bean.

VULGARIS, vul-*gar*-is, common. The Kidney, French and Haricot Beans.

**Philadelphus,** fil-a-*del*-fus; from King Ptolemy Philadelphus. Fragrant flowering shrubs.

CORONARIUS, kor-on-*air*-e-us, crowned or wreathed.

DELAVAYI, del-a-*va*-i, after Delavay, a plant collector.

GORDONIANUS, gor-*don*-c-a-nus, after Gordon.

GRANDIFLORUS, gran-dif-*lo*-rus, large-flowered.

INCANUS, in-*ka*-nus, grey or hoary, the leaf underparts.

INODORUS, in-od-*or*-us, having no fragrance.

MICROPHYLLUS, mi-krof-*il*-lus, small-leaved.

SATSUMI, sat-*su*-mi, Japanese name.

TOMENTOSUS, to-men-*to*-sus, felted—the leaves.

**Philesia,** fil-*e*-ze-a; from Gr. *philein*, to love, in reference to the lily-like flowers. Dwarf shrub.

BUXIFOLIA, buks-if-*o*-le-a, Buxus (box)-leaved.

**Phillyrea,** fil-*ler*-e-a; the ancient Gr. name of the plant. Evergreen shrubs or small trees.

ANGUSTIFOLIA, an-gus-tif-*o*-le-a, narrow-leaved.

LATIFOLIA, lat-if-*o*-le-a, broad-leaved.

**Phlebodium,** fleb-*o*-de-um; from Gr. *phlebs*, a vein, referring to the strong venation or veining of the fronds, and *odous*, a tooth, the shape of the areoles. Warm-house ferns.

AUREUM, *aw*-re-um, golden—the sori and the scales of the rhizomes.

SPORODOCARPUM, spor-od-o-*kar*-pum, spore-fruited—the prominent sori.

**Phlomis,** *flo*-mis; from Gr. *phlomos*, a mullein, or some other woolly plant, which this subject resembled. Herbaceous perennials and shrubs.

FRUTICOSA, frut-ik-*o*-sa, shrubby.

HERBA-VENTI, *herb*-a-*ven*-ti, herb of the wind.

SAMIA, *sa*-me-a, of Samos.

**Phlox,** floks; from Gr. *phlego*, to burn, or *phlox*, a flame, doubtless in allusion to the brightly coloured flowers. Herbaceous perennials and annuals.

AMŒNA, am-*e*-na, lovely, pleasing.

CANADENSIS, kan-a-*den*-sis, of Canada.

DECUSSATA, dek-us-*sa*-ta, divided cross-wise, presumably the leaf arrangement.

DIVARICATA, div-ar-e-*ka*-ta, spreading.

DRUMMONDII, drum-*mun*-dei, after Drummond.

OVATA, o-*va*-ta, egg-shaped.

PANICULATA, pan-ik-ul-*a*-ta, flowers borne in panicles.

PILOSA, pil-*o*-sa, hairy.

STELLARIA, stel-*lair*-e-a, old name; star-shaped flowers.

STOLONIFERA, sto-lon-*if*-er-a, having stolons, or rooted runners.

SUBULATA, sub-u-*la*-ta, awl-like—the leaves.

**Phœnix,** *fe*-niks; the Greek name for the Date Palm, used by Theophrastus. Greenhouse and room palms.

CANARIENSIS, kan-ar-e-*en*-sis, of Canary Islands.

DACTYLIFERA, dak-til-*if*-er-a, date-bearing. The Date Palm.

RŒBELINII, ro-*bel*-in-ei, after Rœbelin.

RUPICOLA, roo-*pik*-o-la, rock loving.

**Phormium,** *for*-me-um; from Gr. *phormos*, a basket, the fibres of the leaves being used for basket-making. Half-hardy perennial.

COOKIANUM, kook-e-*a*-num, after Cook.

TENAX, *te*-naks, tough, the leaf-fibres. The New Zealand Flax.

**Photinia,** fo-*tin*-e-a; from Gr. *photeinos*, shining, referring to the glossy leaves. Evergreen flowering shrubs.

ARBUTIFOLIA, *ar*-but-if-*ol*-e-a, arbutus-leaved.

SERRULATA, ser-rul-*a*-ta, finely toothed.

JAPONICA, jap-*on*-ik-a, of Japan.

**Phragmites,** frag-*my*-teez; from Gr. *phragmos*, a fence, alluding to the fence or hedge-like habit of growth, also Gr. name for reed. Waterside perennial.

COMMUNIS, kom-*mu*-nis, social or common. The Common Reed.

**Phrynium,** *fry*-ne-um; from Gr. *phrynos*, a frog—the plants inhabit marshes. Tropical herbaceous foliage plants.

VARIEGATUM, var-e-eg-*a*-tum, variegated.

**Phuopsis,** fu-*op*-sis; from *Phu*, old Greek name for Valerian, and *opsis*, like, the flowers resembling those of valerians. Rock garden trailer.

STYLOSA, sty-*lo*-sa, the inflorescence having many and prominent styles.

**Phygelius,** fy-*je*-le-us; from Gr. *phugo*, flee, and *helios*, the sun, this plant being a shade lover in its native country. Tender perennial.

CAPENSIS, ka-*pen*-sis, from the Cape of Good Hope.

**Phyllanthus,** fil-*lan*-thus; from Gr. *phyllon*, a leaf, and *anthos*, a flower, the flowers being produced along the edges of the leaf-like phyllodes. Tropical shrubby plants.

GLAUCESCENS, glaw-ses-senz, glaucous or bluish green.

MIMOSOIDES, mim-o-*soy*-dez, mimosa-like.

NIVOSUS, niv-*o*-sus, snowy white—the leaves.

PULCHER, *pul*-ker, beautiful.

ROSEO-PICTUS, ros-e-o-*pik*-tus, rosy-painted—leaves crimson and white.

**Phyllocactus,** fil-lo-*kak*-tus; from Gr. *phyllon*, a leaf, and *cactus*. Greenhouse cacti. Many hybrids.

AKERMANNI, ak-er-*man*-ne, after Akermann.

CRENATUS, kre-*na*-tus, scallop-edged.

**Phyllocactus** (*continued*)

GRANDIS, *gran*-dis, large-flowered.
LATIFRONS, *lat*-if-ronz, broad foliage.
PHYLLANTHOIDES, fil-lanth-*oy*-dez, like P. phyllanthus.

**Phyllodoce**, fil-*lo*-do-ke (more popularly fil-o-*do*-se); name of a sea nymph in the Greek classics. Dwarf shrubs.

EMPETRIFORMIS, em-pet-rif-*or*-mis, like an empetrum, or crowberry.

**Phyllostachys**, fil-*los*-tak-is, from Gr. *phyllon*, a leaf, and *stachys*, a spike, the flowers in leafy spikes. A genus of hardy bamboos.

AUREA, *aw*-re-a, golden, the stems.
NIGRA, *nig*-ra, black, the stems.
VIRIDI-GLAUCESCENS, *ver*-id-e-glaw-*ses*-senz, glaucous-green.

**Phymatodes**, fy-mat-*o*-deez; from Gr. *phymata*, tubercles, the impressed sori (fructification) looking like tubercles on the upper side of the frond. Greenhouse ferns.

ALBO-SQUAMATA, *al*-bo-skwa-*ma*-ta, white scaly—white dots on the fronds.
BILLARDIERI, bil-*lar*-de-air-i, after La Billardiere.
NIGRESCENS, nig-*res*-senz, blackish—dark green fronds.

**Physalis**, *fy*-sa-lis; from Gr. *phusa*, a bladder, in allusion to the inflated calyx. Herbaceous perennials.

ALKEKENGI, al-ke-*ken*-ge, Japanese name. The Bladder Cherry.
FRANCHETTII, fran-*shet*-ei, after Franchett.
IXOCARPA, ix-o-*kar*-pa, viscid—the fruits. The Jamberberry.
PERUVIANA, per-u-ve-*a*-na, of Peru.

**Physostegia**, fy-sos-*te*-je-a; from Gr. *physa*, a bladder, and *stege*, a covering, in reference to the formation of the calyx. Herbaceous perennials.

VIRGINIANA, ver-jin-e-*a*-na, of Virginia.

**Phyteuma**, *fy-tew*-ma; possibly from Gr. *phyteuo*, to plant, or Gr. *phyton*, vegetable growth; used by Dioscorides and adopted by Linnæus. Herbaceous perennials and rock plants.

COMOSUM, kom-*o*-sum, with hairy tufts.
HALLERI, *hal*-ler-i, after Dr. Haller, a botanist.
ORBICULARE, or-bik-ul-*ar*-e, orb-like, the rounded flowers.
SIEBERI, *si*-ber-i, after Sieber, a botanist and plant collector.
SPICATUM, spe-*ka*-tum, the flowers spiked.

**Phytolacca**, fy-tol-*ak*-ka, from Gr. *phyton*, a plant, and *lacca*, lac, in allusion to the crimson colour of the fruit juice. Herbaceous perennials. The Poke Weeds.

ACINOSA, ak-in-*o*-sa, full of kernels.
DECANDRA, dek-*an*-dra, ten-stamened.

**Picea**, *pis*-e-a; ancient L. name. The Spruce.

ALBERTIANA, al-ber-te-*a*-na, of Alberta.
ENGELMANNII, en-gel-*man*-nei, after Engelmann, a botanist of U.S.A.
EXCELSA, ek-*sel*-sa, lofty.
MORINDA, mor-*in*-da, Morinda-like—the cones.
NIGRA, *nig*-ra, black, the dense foliage.
OMORIKA, om-*or*-ik-a, old name for Servian Spruce.
ORIENTALIS, or-e-en-*ta*-lis, eastern.
PUNGENS, *pun*-jenz, sharp—the pointed leaves.
SITCHENSIS, sit-*ken*-sis, of Sitka.
SMITHIANA, smith-e-*a*-na, after Smith.

**Pieris**, *py*-er-is, mythological name. Evergreen shrubs belonging to the heath family. Shrubs.

FLORIBUNDA, flor-ib-*un*-da, free-flowering.
FORMOSA, for-*mo*-sa, beautiful.
JAPONICA, jap-*on*-ik-a, of Japan.

**Pilea**, *py*-le-a; from *pileus*, the Roman felt cap, because of the calyx covering of the achene. Greenhouse foliage perennial.

MICROPHYLLA, mi-krof-*il*-la, small-leaved.
MUSCOSA, mus-*ko*-sa, mossy—a mossy appearance. The Artillery Plant.

**Pilocereus**, py-lo-*se*-re-us, from L. *pilos*, wool, and *cereus*, alluding to the long hairs on the spine cushions. Greenhouse cacti.

SENILIS, sen-*e*-lis, old—the appearance given by the long white hairs. The Old Man Cactus.

**Pilularia**, pil-ul-*a*-re-a; from L. *pilula*, a pill—referring to the round spore cases. Submerged aquatic.

GLOBULIFERA, glob-ul-*if*-er-a, globe-like or round—the spore cases. The Pillwort.

**Pimelea**, py-*meel*-e-a; from Gr. *pimele*, fat, referring to the viscid matter on the leaves of some species and the oily seeds. Greenhouse flowering shrubs.

FERRUGINEA, fer-ru-*jin*-e-a, rusty.
DECUSSATA, dek-us-*sa*-ta, decussate or cross-branched.
ROSEA, *ro*-ze-a, rose-coloured.
SPECTABILIS, spek-*tab*-il-is, showy.

**Pimenta,** py-*men*-ta; from Spanish *pimento*. Warm-house flowering shrubs.

ACRIS, *ak*-ris, acrid.

OFFICINALIS, of-fis-in-*a*-lis, found in shops.

**Pinguicula,** pin-*gwik*-ul-a; from L. *pinguis*, fat or greasy, from the appearance of the leaves in the common Butterwort.

VULGARIS, vul-*gar*-is, common. The Butterwort.

**Pinus,** *py*-nus; ancient classical name for a pine tree. Evergreen coniferous trees.

AUSTRIACA, *aws*-tre-ak-a, of Austria.

CEMBRA, *sem*-bra, old name for the Stone Pine.

COULTERI, *kole*-ter-i, discovered by a Dr. Coulter.

HALEPENSIS, al-ep-*en*-sis, of Aleppo.

LARICIO, lar-*is*-e-o, larch-like; old botanical name.

PINASTER, py-*nas*-ter, old name for the Cluster or Maritime Pine.

PINEA, *py*-ne-a, a pine or Stone Pine. The Umbrella Pine or *Pin Parasol* (Fr.).

PONDEROSA, pon-der-*o*-sa, large and heavy-wooded.

STROBUS, *stro*-bus, old generic name. The Weymouth Pine, *i.e.*, Lord Weymouth's.

SYLVESTRIS, sil-*ves*-tris, of woods. The Scotch Fir or Scot's Pine.

**Piper,** *py*-per; the ancient name. Herbs, woody climbers, shrubs or small trees.

BETLE, *be*-tl, Betel—native name.

EXCELSUM, ek-*sel*-sum, tall or lofty.

NIGRUM, *nig*-rum, black. The Black Pepper.

**Piptanthus,** pip-*tan*-thus; from Gr. *pipto*, to fall, and *anthos*, a flower, alluding to the short duration of the blossoms. Flowering shrubs.

NEPALENSIS, nep-al-*en*-sis, from Nepal.

**Pistia,** *pis*-te-a; from Gr. *pistos*, aquatic. Greenhouse floating aquatic.

STRATIOTES, strat-e-*o*-tez, Stratiotes-like. The Water Lettuce.

**Pisum,** *py*-sum; the classical name, possibly from Celtic *pis*, pea, whence the L. *pisum*, pea. The Garden or Culinary Pea.

SATIVUM, *sat*-iv-um, cultivated.

S. SACCHARATUM, sak-kar-*a*-tum, sugar. The Sugar Pea.

S. UMBELLATUM, um-bel-*la*-tum, umbel-bearing. The Crown Pea.

**Pitcairnia,** pit-*cairn*-e-a; after Archibald Pitcairn, professor of medicine in Scotland and Holland. Tropical flowering perennials.

ANDREANA, an-dre-*a*-na, after Andre.

BRACTEATA, brak-te-*a*-ta, having bracts or coloured floral leaves.

FULGENS, *ful*-jenz, glowing.

**Pittosporum,** pit-*tos*-por-um; from Gr. *pitte*, tar, and *sporos*, seed, the latter being coated with a resinous substance. Half-hardy flowering shrubs.

CRASSIFOLIA, kras-sif-*ol*-e-a, thick-leaved.

TENUIFOLIUM, ten-u-if-*ol*-e-um, thin-leaved.

MAYI, *may*-i, after May.

TOBIRA, to-*bi*-ra, native Japanese name.

UNDULATUM, un-du-*la*-tum, wavy-leaved.

**Plagianthus,** pla-ge-*an*-thus; from Gr. *plagios*, oblique, and *anthos*, a flower, in allusion to the shape of the petals. Flowering trees.

BETULINUS, bet-u-*le*-nus, birch (Betula)-like—the leaves.

LYALLII, *li*-al-ei, after Lyall, botanist and collector.

**Platanus,** *plat*-an-us; the old Greek name for the Plane tree, meaning broad (*platus*), in reference to the palmate leaves or the wide-spreading branches.

ACERIFOLIA, a-ser-if-*o*-le-a, Acer-leaved. The London Plane.

**Platycerium,** plat-e-*ser*-e-um or plat-ik-*eer*-e-um; from Gr. *platys*, broad and *keras*, a horn—in allusion to the broad and horn-shaped fertile fronds. Greenhouse ferns.

ÆTHIOPICUM, eth-e-*o*-pik-um, of Ethiopia (Abyssinia).

ALCICORNE, alk-ik-*orn*-e, elk's-horn. The Elk's-horn Fern.

GRANDE, *grand*-e, large and fine.

**Platyclinis,** plat-ik-*lin*-is, from Gr. *platys*, broad, and *clinis*, a couch, referring to the clinandrium or cavity holding the anthers. Warm-house orchids.

FILIFORMIS, fil-if-*or*-mis, thread-like.

GLUMACEA, glu-*ma*-se-a, large-glumed.

**Platycodon,** plat-ik-*o*-don; from Gr. *platys*, broad, and *kodon*, a bell, in allusion to the large bell-shaped flowers. Herbaceous perennials. The Chinese Bellflower.

GRANDIFLORUM, gran-dif-*lo*-rum, large-flowered.

85

**Platyloma,** plat-il-*o*-ma; from Gr. *platys*, broad, and *loma*, fringe, or edge—the broad sori near the margin. Greenhouse ferns; species as for Pellaea, which see.

**Platystemon,** plat-is-*te*-mon; from Gr. *platys*, broad, and *stemon*, a stamen, in reference to the form of the flowers. Hardy annual.

    CALIFORNICUS, kal-if-*or*-nik-us, of California. The Californian Poppy and Cream Cups.

**Pleione,** pli-*o*-ne; from Gr. mythology, Pleione, mother of the Pleiades. Greenhouse orchids.

    HUMILIS, *hum*-il-is, dwarf.
    LAGENARIA, lag-e-*na*-re-a, bottle-shaped.
    MACULATA, mak-ul-*a*-ta, spotted.
    PRÆCOX    *pra*-koks, precocious, that is, early flowering.

**Pleiospilos,** pla-*os*-pil-os; from Gr. *pleios*, full, and *spilos*, a dot or spot, in allusion to the dotted leaves. Greenhouse succulents.

    BOLUSII, *bo*-lus-ei, after Bolus.
    ROODIÆ, *rood*-e-e, after Mrs. Rood.
    SIMULANS, *sim*-u-lanz, simulating or looking like—in this case, the stony surroundings.

**Pleroma,** ple-*ro*-ma; from Gr. *pleroma*, fullness—the cells in the capsule. Greenhouse flowering shrubs.

    MACRANTHA, mak-*ranth*-a, large-flowered.

**Plumbago,** plum-*ba*-go; from L. *plumbum*, lead, so called by Pliny, who attributed the curing of lead disease to the European species. Greenhouse flowering shrubs. The Leadwort.

    CAPENSIS, ka-*pen*-sis, from Cape of Good Hope.
    LARPENTÆ, lar-*pen*-te, after Lady Larpent.
    ROSEA, *ro*-ze-a, rosy-coloured.
    WILLMOTTIANA, will-mot-te-*a*-na, after Miss Willmott.

**Plumieria,** plu-me-*air*-e-a; after Charles Plumier, a French botanist. Warm-house flowering shrubs.

    ACUTIFOLIA, ak-u-tif-*ol*-e-a, pointed-leaved.
    BICOLOR, *bik*-ol-or, two-coloured.
    RUBRA, *roo*-bra, red. The Frangipanni Plant.

**Podophyllum,** pod-o-*fil*-lum; a contraction of *anapodophyllum*, duck's-foot-leaved—herbaceous perennials.

    EMODI, em-*o*-de, a Himalayan name, of Mount Emodus.
    PELTATUM, pel-*ta*-tum, the leaves shield-shaped.

**Poinciana,** poyn-se-*a*-na; after M. de Poinci, governor of Antilles and a patron of botany. Tropical flowering trees.

    ELATA, e-*la*-ta, high or tall.
    GILLIESII, gil-*lies*-ei, after Gillies.
    REGIA, *re*-je-a, royal.

**Poinsettia,** poyn-*set*-te-a; after a M. Poinsette, a Mexican traveller. Now included in Euphorbia. Tropical perennials.

**Polemonium,** pol-e-*mo*-ne-um; derivation doubtful, possibly Gr. *polemos*, war, from the lance-shaped leaflets, or, as Pliny suggested, because two kings went to war as to which of them had discovered the virtues of the plant. Other authorities state that Dioscorides named it after King Polemon of Pontus. Herbaceous perennials.

    CÆRULEUM, ser-*u*-le-um, sky-blue.
    CARNEUM, *kar*-ne-um, flesh-coloured.
    CONFERTUM, kon-*fer*-tum, crowded—the flowers.
    HUMILE, *hum*-il-e, lowly, dwarf.
    REPTANS, *rep*-tanz, creeping.
    RICHARDSONII, rich-ards-*on*-ei, after Richardson.

**Polianthes,** pol-e-*an*-theez; probably from Gr. meaning white and shining flowers or from Gr. *polis*, a city, and *anthos*, a flower—the use of the flower in city decorations. Greenhouse bulb. The Tuberose.

    TUBEROSA, tu-ber-*o*-sa, bearing tubers.

**Polygala,** pol-*ig*-a-la, from Gr. *polys*, much, and *gala*, milk, the supposition being that the presence of these herbs in pasture increased the production of milk, hence Milkwort—the English name.

    CHAMÆBUXUS, kam-e-*buks*-us, literally false-box, from the resemblance of the leaves to those of box (Buxus).
    MYRTIFOLIA, mer-tif-*o*-le-a, myrtle (Myrtus)-leaved.

**Polygonatum,** pol-ig-on-*a*-tum; from Gr. *polys*, many, and *gonu*, a small joint. Gerarde states that the name alludes to the many knots, or joints, in the roots, and not the stem, as is popularly supposed. Herbaceous perennials.

**Polygonatum** (*continued*)

LATIFOLIUM, lat-if-*o*-le-um, broad-leaved.
MULTIFLORUM, mul-tif-*lo*-rum, many-flowered. Solomon's Seal of gardens.
OFFICINALE, of-fis-in-*a*-le, of the shop, *i.e.*, herbal. The Solomon's Seal.

**Polygonum**, pol-*ig*-on-um; many joints (see under Polygonatum), in reference to the stem formation. Annuals, herbaceous perennials, and climber. The Knotweeds.

AFFINE, af-*fin*-e, kindred, or allied *i.e.*, to some other species.
BALDSCHUANICUM, bawld-shu-*an*-ik-um, of Baldschuania, Bokara.
BISTORTA, bis-*tor*-ta, twice-turned, the twisted root; old generic name for the Bistort.
CAMPANULATUM, kam-pan-u-*la*-tum, bell-flower-like.
COMPACTUM, kom-*pak*-tum, close or compact.
CAPITATUM, kap-it-*a*-tum, flowers in heads.
POLYSTACHYUM, pol-is-*tak*-e-um, many-spiked.
SACHALINENSE, sak-al-in-*en*-se, of Sakhalin Island.
VACCINIFOLIUM, vak - sin - if-*o* - le - um, leaved like vaccinium.

**Polypodium**, pol-e-*po*-de-um; from Gr. *polys*, many, and *pous*, a foot (many little feet), in allusion to the furry foot-like divisions of the creeping stems. Hardy and greenhouse evergeeen ferns.

ALBO-SQUAMATUM, *al*-bo-skwa-*ma*-tum, white scaly—dots on the fronds.
ALPESTRE, al-pes-tre, of mountains.
AUREUM, *aw*-re-um, golden—the sori or rhizome scales.
BILLARDIERI, bil-lar-de-*air*-i, after La Billardiere.
DRYOPTERIS, dry-*o*-ter-is, from Gr. *drys*, oak, and *pteris*, fern. The Oak Fern.
PHEGOPTERIS, feg-*o*-ter-is, from *phegos*, beech, and *pteris*, fern. The Beech Fern.
PUSTULATA, pus-tu-*la*-ta, covered with bladder-like excrescences—the seats of the sori.
VULGARE, vul-*gar*-e, common. The Polypody.
v. CAMBRICUM, *cam*-bre-kum, of Wales. The Welsh Polypody.

**Polystichum**, pol-*is*-tik-um; from Gr. *polys*, many, and *stichos*, a row, in reference to the several rows of spore-cases. Greenhouse and hardy ferns.

ACROSTICHOIDES, ak-*ros*-tik-*oy*-dez, acrostichum-like.

**Polystichum** (*continued*)

ACULEATUM, ak-u-le-*a*-tum, prickly—the acute pinnæ. The Prickly Shield Fern.
ANGULARE, ang-ul-*a*-re, angular—the pinnæ. The Soft Shield Fern.
LONCHITIS, lon-*ki*-tis, spear-like. The Holly Fern.
MUNITUM, mu-*nee*-tum, armed.

**Pontederia**, pon-te-*deer*-e-a; named after J. Pontedera, once professor of botany at Padua. Hardy aquatic.

CORDATA, kor-*da*-ta, heart-shaped—the leaves.

**Populus**, *pop*-u-lus; the *arbor-populi* (tree of the people) of the Romans, the Italian or Lombardy Poplar being much planted in their cities.

ALBA, *al*-ba, white. The White Poplar.
BALSAMIFERA, bawl-sam-*if*-er-a, balsam-scented.
CANDICANS, *kan*-dik-ans, shining white —the leaves.
CANESCENS, kan-*es*-sens, greyish-white, hoary.
MONILIFERA, mon-il-*if*-er-a, necklace bearing—the long fruits.
NIGRA, *nig*-ra, black.
N. ITALICA, it-*al*-ik-a, of Italy. The Lombardy Poplar.
SEROTINA, ser-o-*tin*-a, late in starting spring growth.
TREMULA, *trem*-ul-a, trembling, the quivering of the leaves. The Aspen.
TRICHOCARPA, trik - o - *kar* - pa, hairy-fruited.

**Portulaca**, por-tu-*lak*-a; an old L. name, possibly from L. *porto*, to carry, and *lac*, milk, alluding to the milky juice. Annuals.

GRANDIFLORA, gran-dif-*lo*-ra, large-flowered.
OLERACEA, o-ler-*a*-se-a, a pot-herb. The Purslane.

**Posoqueria**, po-zo-*ke*-re-a; part of the native Guianan name—Aymara poso-queri. Tropical flowering shrubs.

FORMOSA, for-*mo*-sa, beautiful.
FRAGRANTISSIMA, fra-gran-*tis*-sim-a, very fragrant.
LONGIFLORA, long-if-*lo*-ra, long-flowered.

**Potamogeton**, pot-a-mog-*e*-ton; from Gr. *potamos*, a river, and *geiton*, a neighbour, growing in rivers or ponds. Submerged aquatics. The Pondweeds.

CRISPUS, *kris*-pus, curled—wavy-edged leaves.
LUCENS, lu-*senz*, shining—the foliage.

**Potamogeton** (*continued*)

PECTINATUS, pek-tin-*a*-tus, feathered or comb-like—the arrangement of the leaves.

PERFOLIATUS, per-fol-e-*a*-tus, perfoliate or stem-clasping leaves.

PUSILLUS, pu-*sil*-lus, small—the plant.

**Potentilla**, po-ten-*til*-la; from L. *potens*, powerful, some species having active medicinal properties. Rock plants, herbaceous perennials, and shrubs.

ARGENTEA, ar-*jen*-te-a, silvery, the leaves.

ARGYROPHYLLA, ar-ger-*of*-il-la, silvery-leaved.

AUREA, *aw*-re-a, golden—the flowers.

DAVURICA, da-*voo*-rik-a, of Davuria, Siberia.

FRAGIFORMIS, fraj-if-*or*-mis, strawberry-like.

FRUTICOSA, frut-ik-*o*-sa, shrubby.

NEPALENSIS, nep-al-*en*-sis, of Nepal.

NITIDA, *nit*-id-a, smooth and lustrous.

PURDOMII, *pur*-dom-ei, after Purdom.

RUPESTRIS, roo-*pes*-tris, of rocks.

TRIDENTATA, trid-en-*ta*-ta, three-lobed, the leaves.

**Poterium**, po-*teer*-e-um; from Gr. *poterion*, a drinking-cup, the shape of the calyx in these herbs. The Burnets.

CANADENSE, kan-a-*den*-se, Canadian.

OBTUSUM, ob-*tu*-sum, blunt or rounded.

SANGUISORBA, san-gwis-*or*-ba, old generic name for this plant. The Salad Burnet.

**Pratia**, *pra*-te-a; after Prat-Bernon, a French naval officer. Trailing perennials.

ANGULATA, ang-ul-*a*-ta, angled, presumably the growths.

BEGONIFOLIA, be-go-nif-*o*-le-a, begonia-leaved.

**Primula**, *prim*-u-la (or *pri*-mul-a); from L. *primus*, first, referring to the early flowering of many of the Primroses. Hardy and greenhouse herbaceous plants, some treated as annuals.

ACAULIS, a-*kaw*-lis, stemless. The Primrose.

AURICULA, aur-*ik*-ul-a, from L. *auricula*, an ear—the ear-like leaves. The Auricula.

BEESIANA, beez-e-*a*-na, after Messrs. Bees, Ltd., the nurserymen.

BULLEYANA, bul-le-*a*-na, after Mr. A. K. Bulley.

CAPITATA, kap-it-*a*-ta, flowers in a head.

CASHMERIANA, kash-mer-e-*a*-na, of Kashmir.

CHIONANTHA, ki-on-*an*-tha, snow-white flowers.

**Primula** (*continued*)

COCKBURNIANA, kok-burn-e-*a*-na, after Cockburn.

DENTICULATA, den-tik-u-*la*-ta, finely toothed—the leaves.

ELATIOR, e-*la*-te-or, taller. The Oxlip.

FARINOSA, far-in-*o*-sa, mealy, the foliage and stems.

FLORINDÆ, flor-*in*-de, after Florind.

FORRESTII, for-*res*-tei, after Forrest, plant collector.

FRONDOSA, fron-*do*-za, leafy.

GLAUCESCENS, *glaw*-ses-senz, somewhat glaucous.

GLYCOSMA, gli-*kos*-ma, sweet-scented, the leaves.

HELODOXA, hel-o-*doks*-a, the glory of the bog.

HIRSUTA, hir-*su*-ta, hairy.

INVOLUCRATA, in-vol-u-*kra*-ta, the flowers ruffed, or involucred.

JAPONICA, jap-*on*-ik-a, of Japan.

JULIÆ, *ju*-le-e, after Julia.

LITTONIANA, lit-to-ne-*a*-na, after Litton, plant collector.

MALACOIDES, mal-ak-*oy*-dez, mallow-like, presumably the colour.

MARGINATA, mar-jin-*a*-ta, margined (with white)—the leaves.

MICRODONTA, mi-krod-*on*-ta, small-toothed—the leaves.

OBCONICA, ob-*kon*-ik-a, an inverted cone —the calyx.

PULVERULENTA, pul-ver-ul-*en*-ta, powdered with meal.

ROSEA, *ro*-ze-a, rosy.

SAXATILIS, saks-*a*-til-is, rock-haunting.

SECUNDIFLORA, sek-un-dif-*lo*-ra, the flowers of the cluster all turned one way.

SIEBOLDII, se-*bold*-ei, after Siebold, German botanist and traveller.

SINENSIS, sin-*en*-sis, of China. The Chinese Primula.

SIKKIMENSIS, sik-kim-*en*-sis, of Sikkim.

VARIABILIS, var-e-*ab*-il-is, variable. The Polyanthus.

VEITCHII, *veech*-ei, after Veitch, the celebrated nurseryman.

VILLOSA, vil-*lo*-sa, shaggy, the leaves hairy.

VISCOSA, vis-*ko*-sa, clammy or gummy.

VULGARIS, vul-*gar*-is, common. The Primrose.

WINTERI, *win*-ter-i, after Winter.

**Prostanthera**, pros-tan-*the*-ra; from Gr. *prostheke*, appendage, and *anthera*, an anther—the connectives of the anthers are spurred. Tender flowering shrub.

ROTUNDIFOLIA, ro-tun-dif-*o*-le-a, round-leaved.

**Prumnopitys,** prum-*nop*-it-is; from Gr. *prumnos,* the last or extreme, and *pitys,* a pine. Hardy conifer. The Plum Fir.

ELEGANS, *el*-e-ganz, elegant—the habit.

**Prunella,** proo-*nel*-la; from German *braume,* quinsey, which the plant was supposed to heal. Also spelt Brunella, which may be correct. Rock and border perennials.

GRANDIFLORA, gran-dif-*lo*-ra, large-flowered.

VULGARIS, vul-*gar*-is, common.

**Prunus,** *proo*-nus; classical name of the plum. Flowering and fruiting trees.

AMYGDALORA, am-*ig*-da-lus. See "Amygdalus."

AVIUM, *av*-e-um, L. *avis,* a bird. The Common Gean, sometimes called Bird-Cherry.

CERASIFERA, ser-as-*if*-er-a, cherry-bearing. The Cherry-Plum.

CERASUS, ser-*a*-sus, Gr. *kerasos,* a Cherry tree, said to have come from Cerasus in Pontus. The Cherry Tree.

COMMUNIS, kom-*mu*-nis, common, *i.e.,* in groups or communities. The Almond.

CORNUTA, kor-*nu*-ta, horned, the shape of the fruits.

DAVIDIANA, da-vid-e-*a*-na, after Abbé David.

INCANA, in-*ka*-na, woolly, the leaf undersides.

LAUROCERASUS, *law*-ro-ser-*a*-sus, laurel-cherry. The Cherry Laurel or "Laurel" of gardens.

LUSITANICA, loo-sit-*a*-nik-a, Portuguese. The Portugal Laurel.

MAHALEB, ma-*a*-leb, a Arabic name.

MUME, *mu*-me, a Japanese name.

PADUS, *pa*-dus, Gr. name for the true Bird-Cherry.

PERSICA, *per*-sik-a, of Persia. The Peach and Nectarine (see "Persica").

SERRULATA, ser-rul-*a*-ta, finely toothed —the leaves.

SPINOSA, spin-*o*-sa, spiny.

SUBHIRTELLA, sub-hir-*tel*-la, slightly hairy—the leaves and young wood.

TRILOBA, tril-*o*-ba, the leaves usually three-lobed.

**Pseudotsuga,** *sue*-do-*su*-ga; the false Tsuga, *i.e.,* allied to, the true Tsugas (Hemlock Firs). This group now only comprises the Douglas Fir.

DOUGLASII, dug-*las*-ei, after Douglas.

**Psidium,** *sid*-e-um; from Gr. *psidion,* the Greek name for pomegranate. Tropical fruiting tree. The Guava.

GUAVA, gu-*av*-a, The Guava.

**Ptelea,** *tel*-e-a; from Gr. *ptelea,* the Greek name for elm, the similarity residing in the winged fruits. Ornamental tree.

TRIFOLIATA, trif-o-le-*a*-ta, three-leaved.

**Pteris,** *ter*-is; from Gr. *pteron,* a wing, the branched frond of the Bracken Fern resembling a pair of outspread wings. Greenhouse ferns.

AQUILINA, ak-wil-*e*-na, an eagle, various explanations offered, most probable being that suggested above. The Bracken Fern.

CRETICA, *kret*-ik-a, of Crete.

C. ALBO-LINEATA, *al*-bo-lin-e-*a*-ta, white lined—variegated with white.

LONGIFOLIA, long-if-*ol*-e-a, long-leaved.

PALMATA, pal-*ma*-ta, hand-shaped.

SERRULATA, ser-rul-*a*-ta, finely toothed.

TREMULA, *trem*-ul-a, trembling.

UMBROSA, um-*bro*-za, growing in shady places.

WIMSETTII, wim-*set*-tei, after Wimsett.

**Pterocactus,** ter-o-*kak*-tus; from Gr. *pteron,* a wing, and *cactus,* reference being to the winged seeds. Greenhouse cacti.

PUMILUS, *pu*-mil-us, dwarf or small.

TUBEROSUS, tu-ber-*o*-sus, bearing tubers.

**Pulmonaria,** pul-mon-*air*-e-a; from L. *pulmo,* pertaining to the lungs, one species having been regarded as a remedy for diseases of the lungs, hence Lungwort. Rock garden and woodland perennials.

ANGUSTIFOLIA, an-gus-tif-*o*-le-a, narrow-leaved.

OFFICINALIS, of-fis-in-*a*-lis, of the shop (herbal). The Lungwort.

RUBRA, *roo*-bra, red—the flowers.

SACCHARATA, sak-kar-*a*-ta, sugared; application obscure, but possibly alluding to the white-powdered leaves.

**Punica,** *pu*-ne-ka; from Malum punicum, Apple of Carthage, an early name for Pomegranate. Greenhouse flowering wall shrub.

GRANATUM, gra-*na*-tum, old substantive name.

**Puschkinia,** poos-*kin*-e-a; after M. Pouschkin, a Russian botanist. Half-hardy bulbs.

SCILLOIDES, sil-*oy*-dez, scilla-like.

**Pyracantha,** per-a-*kan*-tha; from Gr. *pyr,* fire, and *akanthos,* a thorn, probably in reference to the brilliant berries and spiny branches. The Fire-Thorn.

**Pyracantha** (*continued*)

ANGUSTIFOLIA, an-gust-tif-*o*-le-a, narrow-leaved.

COCCINEA, kok-*sin*-e-a, scarlet—the fruits.

GIBBSII, *gibbs*-ei, after Gibbs.

LALANDEI, la-*lan*-de-i, after La Lande.

**Pyrethrum**, py-*re*-thrum; from Gr. *pyr*, fire, probably fever-heat, since the plant was used in ancient medicine to assuage fevers. Rock and herbaceous perennials.

PARTHENIUM, par-*then*-e-um, common-pellitory. The Feverfew.

P. AUREUM, *aw*-re-um, golden—the leaves. The Golden Feather of summer bedding.

ROSEUM, ro-ze-um, rosy. The Garden Pyrethrum in double and single varieties.

**Pyrola**, *py*-ro-la; from L. *pyrus*, a Pear tree, application not obvious unless it refers to the shape of the leaves. Dwarf rock and woodland plants.

ROTUNDIFOLIA, ro-tun-dif-*o*-le-a, round-leaved.

**Pyrus**, *py*-rus; classical name for a Pear tree. Fruiting and flowering trees. The genus now includes Aria (Whitebeam), Malus (Crab), Sorbus (Mountain Ash), and others.

AMYGDALIFORMIS, am-ig-dal-if-*or*-mis, almond-like.

ARIA, *a*-re-a, old generic term for Whitebeams, probably Persian place-name of origin.

AUCUPARIA, aw-ku-*par*-e-a, old name for Mountain Ash, from L. *aucupium*, bird-catching, from the ancient belief that the berries intoxicated birds, rendering them easily caught. The Rowan Tree or Mountain Ash.

BACCATA, bak-*ka*-ta, berried. The Cherry Apple or Siberian Crab.

CHAMÆMESPILUS, kam-e-*mes*-pil-us, lit. ground medlar—its dwarf stature.

CORONARIA, kor-on-*air*-e-a, crowned—the shape of the tree.

ELEYI, *ee*-ley-i, after Charles Eley, the raiser.

FLORIBUNDA, flor-ib-*un*-da, free-flowering.

MALUS, *ma*-lus, L. for Apple tree; old generic name. The Crab Apple.

MELANOCARPA, mel-an-ok-*ar*-pa, black-fruited.

PRUNIFOLIA, proo-nif-*o*-le-a, plum-tree (Prunus)-leaved.

**Pyrus** (*continued*)

SCHIEDECKERI, shi-*dek*-er-i, after Schiedecker, a botanist.

SIKKIMENSIS, sik-kim-*en*-sis, of Sikkim.

SINENSIS, si-*nen*-sis, Chinese.

SORBUS, *sor*-bus, from L. *sorbum*, (Sorb Apple), once the generic name. The Service Tree.

SPECTABILIS, spek-*tab*-il-is, showy.

TORMINALIS, tor-min-*a*-lis, against colic. The Wild Service Tree.

TORINGO, tor-*in*-go, a Japanese name.

VILMORINII, vil-mor-*e*-nei, after M. Maurice de Vilmorin, who raised it.

**Quercus**, *kwer*-kus; Latin name for an Oak tree; some authorities derive word from Celtic, *quer*, fine, and *cuez*, a tree. Evergreen and deciduous trees.

CASTANEÆFOLIA, kas-tan-e-e-*fo*-le-a, chestnut (Castanea)-leaved.

CERRIS, *ser*-ris, old (Lat.) name for this tree. The Turkey Oak.

COCCIFERA, kok-*sif*-er-a, coccus-bearing, here alluding to the kermes insect parasitic on this tree, and which yields a scarlet dye, hence Kermes Oak.

COCCINEA, kok-*sin*-e-a, scarlet—the autumnal colouring of foliage. The Scarlet Oak.

ILEX, *i*-leks, ancient Lat. name for the Holm Oak, now generic name for the hollies which some evergreen oaks resemble. The Holm, Evergreen or Holly-leaved Oak.

LIBANI, *lib*-an-e, of Lebanon.

LUCOMBEANA, luk-om-be-*a*-na, Lucombe. The Lucombe Oak.

PEDUNCULATA, ped-ungk-ul-*a*-ta, long-stalked, the inflorescences and fruits. The English Oak.

PHELLOS, *fel*-los, willow. The Willow Oak.

PONTICA, *pon*-tik-a, Pontic, the shores of the Black Sea.

RUBRA, *roo*-bra, red—the autumn leaf colour.

SESSILIFLORA, ses-sil-if-*lo*-ra, flowers and fruit stalkless, or nearly so.

SUBER, *su*-ber, cork, old Lat. name for this tree. The Cork Oak.

VELUTINA, vel-u-*te*-na, velvety—young wood and buds downy.

**Ramonda**, ra-*mon*-da; after von Ramond de Carbonnieres, a French botanist. Rock plants.

SERBICA, *ser*-bik-a, of Serbia.

PYRENAICA, pir-en-*a*-ik-a, of Pyrenees.

**Ranunculus**, ra-*nun*-kul-us, from L. *rana*, a frog, some species inhabiting marshy places where frogs abound. Herbaceous, waterside, and rock plants.

**Ranunculus** (*continued*)

ACONITIFOLIUS, ak-on-e-tif-*o*-le-us, leaved like aconite (Aconitum).

ACRIS, *ak*-ris, sharp or bitter. The Meadow Crowfoot or Buttercup. A double variety is grown in gardens.

AMPLEXICAULIS, am - pleks - e - *kaw* - lis, leaves stem clasping.

ANEMONOIDES, an-em-on-*oy*-dez, anemone-like.

ASIATICUS, a-she-*at*-ik-us, Asian. The Persian Ranunculus.

CRENATUS, kre-*na*-tus, leaves crenated or scalloped.

GLACIALIS, glas-e-*a*-lis, icy; a high alpine plant.

GRAMINEUS, gram-*in*-e-us, grassy—the leaves.

LINGUA, *ling*-u-a, a tongue—the shape of the leaves.

LYALLII, ly-*al*-lei, after Lyall. The Rockwood Lily.

NEMOROSUS, nem-or-*o*-sus, pertaining to groves. The Wood Anemone.

NIVALIS, niv-*a*-lis, snowy, or lofty regions.

NYSSANUS, nis-*sa*-nus, from Nyssa.

PARNASSIFOLIUS, par-nas-sif-*o*-le-us, leaved like Parnassia.

RUTÆFOLIUS, ru-ta-*fo*-le-us, rue (Ruta)-leaved.

**Raoulia,** ra-*oo*-le-a; after E. Raoul, a French naval surgeon who wrote on New Zealand plants when in New Zealand waters. Mat-forming rock plants.

AUSTRALIS, aws-*tra*-lis, southern.

GLABRA, *gla*-bra, smooth, or without hairs.

SUBSERICIA, sub-ser-*is*-e-a, somewhat downy.

**Raphanus,** *raf*-an-us; classical name, possibly from Gr. *ra*, quickly, and *phaino*, I appear, in reference to the quick germination of the seeds.

CAUDATUS, kaw-*da*-tus, tailed. The Rat-tailed Radish.

SATIVUS, *sat*-iv-us, cultivated. The Radish.

**Raphiolepis,** raf-e-o-*le*-pis; from Gr. *raphis*, a needle, and *lepis*, a scale, probably in allusion to the formation of the bracts. Shrubs.

INDICA, *in*-dik-a, of India.

OVATA, o-*va*-ta, egg-shaped—the leaves.

**Rebutia,** re-*but*-e-a; after P. Rebut, a trader in cacti. Greenhouse cacti.

GRANDIFLORA, gran-dif-*lo*-ra, large-flowered.

MINUSCULA, min-us-*ku*-la, small—the plant.

**Rehmannia,** ra-*man*-ne-a; named after J. Rehmann, a Russian doctor. Greenhouse perennials.

ANGULATA, ang-ul-*a*-ta, angled—the stems.

ELATA, e-*la*-ta, tall.

HENRYI, *hen*-re-i, after Dr. Henry.

**Reinwardtia,** rin-*wardt*-e-a; after K. G. K. Reinwardt, director of Leyden Botanic Garden. Greenhouse flowering shrubs.

TETRAGYNUM, tet-*ra*-jin-um, four-styled.

TRIGYNUM, *trig*-in-um, three-styled.

**Reseda,** re-se-da; from L. *resedo*, to heal, or assuage, the name being given by Pliny to a species of mignonette which was believed to possess certain medicinal virtues. Hardy annual.

ODORATA, od-o-*ra*-ta, scented. The Mignonette.

**Retinospora,** ret-in-*os*-por-a; from Gr. *retine*, resin, and *spora*, seed (yielding). Coniferous trees.

OBTUSA, ob-*tu*-sa, obtuse or blunt.

PISIFERA, pis-*if*-er-a. Pisum or pea-bearing—the round fruits or cones.

PLUMOSA, plu-*mo*-sa, feathery.

**Rhamnus,** *ram*-nus; ancient Gr. name, possibly from Gr. *rhamnos*, name for a thorny shrub. Deciduous or evergreen trees or shrubs.

ALATERNUS, al-a-*ter*-nus, an old generic name of doubtful application.

CATHARTICA, kath-*ar*-tik-a, cathartic. The Buckthorn.

COSTATA, kos-*ta*-ta, leaves conspicuously ribbed.

FRANGULA, *frang*-ul-a, brittle; old name for this buckthorn.

INFECTORIA, in-fec-*tor*-e-a, yielding dye.

PURSHIANA, pur-she-*a*-na, after Pursh, a German botanist.

**Rhapis,** *rap*-is; from Gr. *rhaphis*, a needle, possibly referring to the sharply pointed leaves, or to the acute awns of the corolla. Greenhouse palms. The Ground Rattan Cane.

FLABELLIFORMIS, fla-bel-lif-*or*-mis, fan-shaped—the leaves.

HUMILIS, *hum*-il-is, humble or dwarf.

**Rheum,** *re*-um; from *Rha*, the Russian name of the River Volga, near which rhubarb was first found. The old Greeks called rhubarb *Rha*; the Eng. name of Rhubarb for the edible plant (R. rhaponticum) has the same origin. Herbaceous perennials.

**Rheum** (*continued*)

EMODI, em-*o*-de, Mt. Emodus, Himalayas.

OFFICINALE, of-fis-in-*a*-le, of the shop (herbal).

PALMATUM, pal-*ma*-tum, palmate (hand)-leaved.

RHAPONTICUM, ra-*pon*-tik-um, Pontic Rha. The Culinary Rhubarb.

**Rhipsalis**, *rip*-sa-lis; from Gr. *rhips*, a willow or wicker work, referring to the slender interlacing branches. Greenhouse cacti.

CASSYTHA, kas-*sy*-tha, meaning not known.

CRISPATA, kris-*pa*-ta, curled.

HOULLETII, *howl*-let-ei, after Houllet.

PARADOXA, par-a-*dox*-a, paradoxical—the appearance of the plant.

**Rhodanthe**, ro-*dan*-the; from Gr. *rhodon*, a rose, and *anthos*, a flower, in reference to the deep red colour. Half-hardy annuals.

MACULATA, mak-ul-*a*-ta, blotched.

MANGLESII, mang-*le*-sei, after Captain Mangles.

**Rhodochiton**, ro-do-*ky*-ton or rod-*ok*-it-on; from Gr. *rhodo*, rose, and *chiton*, a cloak—the calyx is swollen and red-coloured. Greenhouse flowering climber.

VOLUBILE, vol-*u*-bil-e, twisting round.

**Rhododendron**, ro-do-*den*-dron; from Gr. *rodon*, a rose, and *dendron*, a tree, the "Alpine Rose" or *Rose des Alpes* (R. ferrugineum), probably being the first true rhododendron to receive that name.

AMBIGUUM, am-*big*-u-um, uncertain, possibly alluding to the indefinite colour.

AMŒNUM, am-*e*-num, pleasing, lovely.

ARBORESCENS, ar-bor-*es*-senz, tree-like.

ARBOREUM, ar-*bor*-e-um, tree-like.

AUGUSTINI, aw-gus-*te*-ne, after Dr. Augustine Henry.

AURICULATUM, aw-rik-ul-*a*-tum, the leaves auricled (ear-shaped).

BALSAMINÆFLORA, bawl-*sam*-in-e-*flo*-ra, balsam-flowered—the double-flowered florists' balsam.

CALENDULACEUM, kal-en-du-*la*-se-um, resembling a marigold—the brilliant colour.

CALOPHYTUM, kal-o-*fy*-tum, lit. beautiful plant.

CALOSTROTUM, kal-os-*tro*-tum, lit. beautiful covering, presumably the silvered leaves.

CAMPANULATUM, kam-pan-u-*la*-tum, bell-shaped flowers.

CAMPYLOCARPUM, kam-pil-o-*kar*-pum, bearing bent fruits.

**Rhododendron** (*continued*)

CAMPYLOGYNUM, kam-pil-o-*jin*-um, the style curved.

CATAWBIENSE, kat-aw-be-*en*-se, of Catawba, U.S.A.

CAUCASICUM, kaw-*kas*-ik-um, Caucasian.

CILIATUM, sil-e-*a*-tum, fringed with bristles—the leaves.

CINNABARINUM, sin-na-bar-*e*-num, cinnabar-red.

DAURICUM, *daw*-rik-um, of Dahuria, Siberia.

DECORUM, dek-*or*-um, shapely or becoming.

DICHROANTHUM, dik-ro-*an*-thum, with bicoloured flowers.

DISCOLOR, dis-*ko*-lor, variously coloured flowers.

FALCONERI, fal-*kon*-er-i, after Dr. Falconer.

FASTIGIATUM, fas-tij-i-*a*-tum, fastigiate—erect branches tapering to a point.

FERRUGINEUM, fer-ru-*jin*-e-um, rusty—the leaf undersides. The Alpine Rose.

FLAVUM, *fla*-vum, yellow.

GLAUCUM, *glaw*-kum, glaucous—or blue-green—the leaf undersides.

GRIFFITHIANUM, grif-fith-e-*a*-num, after Griffith.

HÆMATODES, hem-a-*to*-dez, blood-like—colour of the flower.

HIPPOPHÆOIDES, hip-po-fa-*oy*-dez, resembling Sea Buckthorn (Hippophæ).

IMPEDITUM, im-ped-*e*-tum, tangled—the twiggy branches.

INDICUM, *in*-dik-um, of India. The type of evergreen large-flowered "azalea" grown in greenhouses.

INTRICATUM, in-trik-*a*-tum, webby—the scurfy covering.

LAPPONICUM, lap-*pon*-ik-um, of Lapland.

LEDIFOLIUM, led-e-*fo*-le-um, Ledum-leaved.

LEDOIDES, led-*oy*-dez, Ledum-like.

LEPIDOTUM, lep-id-*o*-tum, beset with scales.

MOLLE, *mol*-le, soft or velvety—the leaves.

MOUPINENSE, moo-pin-*en*-se, of Moupin, W. Szechuan.

MYRTILLOIDES, mir-til-*oy*-dez, myrtle-like.

NERIIFLORUM, ne-re-e-*flo*-rum, Nerium (oleander)-flowered.

NUDIFLORUM, nu-dif-*lo*-rum, naked-flowered, *i.e.*, coming before the leaves.

OLEIFOLIUM, ol-e-if-*ol*-e-um, Olea (olive)-leaved.

ORBICULARE, or-bik-ul-*ar*-e, orbicular or round—the leaves.

OREOTREPHES, or-e-o-*tre*-fez, mountain dweller.

PONTICUM, *pon*-tik-um, Pontic, region of Black Sea.

**Rhododendron** (*continued*)

PRÆCOX, *pra*-koks, early—as to flowering.
PRIMULINUM, prim-ul-*e*-num, primrose-like—the flowers.
PUNCTATUM, pungk-*ta*-tum, dotted—the leaves.
RACEMOSUM, ra-se-*mo*-sum, flowers in raceme-like clusters.
RHODORA, ro-*dor*-a, old generic name signifying rosy-red.
RUSSATUM, rus-*sa*-tum, reddened—the foliage.
SALUENSE, sal-u-*en*-se, from Salween, W. Szechuan.
SCHLIPPENBACHII, schlip-pen-*bach*-ei, after one of its discoverers, Baron Schlippenbach.
SCINTILLANS, *sin*-til-lanz, sparkling, bright.
SERPYLLIFOLIUM, ser - pil - lif - *o* - le - um, leaved like thyme (Thymus serpyllum).
SOULIEI, *soo*-le-i, discovered by Soulie.
SUTCHUENSE, sut-ku-*en*-se, from Szechuan.
THOMSONII, tom-*son*-ei, after Thomson.
VASEYI, va-*ze*-i, discovered by Mr. G. R. Vasey.
VISCOSUM, vis-*ko*-sum, sticky or viscid.
WILLIAMSIANUM, wil-yamz-e-*a*-num, after Williams.
YUNNANENSE, yun-nan-*en*-se, of Yunnan, S. China.

**Rhodora,** ro-*dor*-a; from Gr. *rhodon*, a rose, the colour of the blossom. Flowering shrub.

CANADENSIS, kan-a-*den*-sis, of Canada. Now Rhododendron Rhodora.

**Rhodothamnus,** ro-do-*tham*-nus; Gr. *rhodon*, a rose, and *thamnos*, a bush or shrub—the flowers rose-coloured. Flowering shrub.

CHAMÆCISTUS, kam-e-*sis*-tus, Gr. *chamai*, on the ground, *kistos*, a rock-rose, an old generic name of misleading application (Ground Cistus), this dwarf shrub being ericaceous.

**Rhodotypos,** ro-*do*-tip-os; from Gr. *rhodon*, a rose, and *tupos*, shape or form, flowers suggesting a small white rose. Shrub.

KERRIOIDES, ker-re-*oy*-dez, like Kerria. The White Kerria.

**Rhoeo,** *ro*-e-o; the author of this name (Hance) has given no account of its meaning. Greenhouse trailers.

DISCOLOR, *dis*-kul-or, flowers variously or part-coloured.

**Rhus,** roos; ancient Gr. name. Foliage shrubs with milky or resinous juice.

COPALLINA, kop-al-*le*-na, yielding copal (lacquer).
COTINOIDES, kot-in-*oy*-dez, like Cotinus.
COTINUS, *kot*-in-us, old generic name for this Sumach, signifying a wild olive. The Smoke Bush or Wig Tree.
GLABRA, *gla*-bra, smooth—the leaves.
TOXICODENDRON, toks-ik-od-*en*-dron, poison tree. The Poison Ivy.
TYPHINA, *ty*-fin-a, antler-shaped—the branches. The Stag's-horn Sumach.
VERNICIFERA, ver-nik-*if*-er-a, varnish-yielding.

**Rhynchospermum,** rin-kos-*per*-mum; possibly from Gr. *rhyncos*, a snout or beak, and *sperma*, a seed. Greenhouse climbing shrub.

JASMINOIDES, jas-min-*oy*-dez, jessamine (Jasminum)-like.

**Ribes,** *ry*-bees; origin uncertain, possibly the Arabic *Ribas*, an acid plant used by Arabian physicians and known to science as Rheum Ribes. Flowering and fruiting shrubs.

AUREUM, *aw*-re-um, golden—the flowers.
GROSSULARIA, gros-sul-*a*-re-a, rough. The Gooseberry.
NIGRUM, *ni*-grum, black—the fruit. The Black Currant.
SANGUINEUM, sang-*win*-e-um, blood-red—the flowers.
SPECIOSUM, spes-e-*o*-sum, showy. The Fuchsia Currant.
VULGARE, vul-*gar*-e, common. The Red Currant.
V. ALBUM, *al*-bum, white. The White Currant.
VIBURNIFOLIUM, vi-bur-nif-*o*-le-um, Viburnum-leaved.

**Riccia,** *rich*-e-a; after Pietro Francisco Ricci, a Florentine botanist. Greenhouse floating cryptogamic plants known as Crystal-worts.

FLUITANS, *flu*-it-anz, floating.

**Richardia,** rik-*ar*-de-a; named after R. Richardson, English physician. Greenhouse herbaceous perennials.

ÆTHIOPICA, eth-i-*o*-pik-a, Ethiopian.
AFRICANA, af-rik-*a*-na, African—South African.
ALBO-MACULATA, *al*-bo-mak-ul-*a*-ta, white spotted—the leaves.
ELLIOTTIANA, el-le-ot-e-*a*-na, after Elliott.
PENTLANDII, *pent*-land-ei, after Pentland.
REHMANNII, ra-*man*-nei, after Rehmann

**Ricinus,** *ris*-in-us; classical L. name, possibly from L. *ricinus*, a tick or bug which the seed resembles. Foliage annuals.

COMMUNIS, kom-*mu*-nis, common or social. The Castor Oil Plant.

**Rimaria,** rim-a-*re*-a; from L. *rima*, cleft or fissure, in allusion to the slit between the close-pressed leaves when the plants are dormant. Greenhouse succulents.

HEATHII, *heeth*-ei, after Dr. Rodier Heath.

ROODLÆ, *rood*-e-e, after Mrs. E. Rood.

**Rivina,** riv-*ee*-na; after A. Q. Rivinus, professor of botany in Leipzig. Warmhouse fruiting shrub.

HUMILIS, *hum*-il-is, low or dwarf.

**Robinia,** rob-*in*-e-a; named after Jean Robin, a French botanist and herbalist to Henry IV of France, and Vespasien Robin, his son. Shrubs and trees.

HISPIDA, *his*-pid-a, bristly—the twigs.

KELSEYI, kel-*se*-i, after Kelsey, nurseryman, Boston, U.S.A.

PSEUDO-ACACIA, *sue*-da-*ka*-she-a, false acacia.

P. INERMIS, in-*er*-mis, unarmed, *i.e.*, not thorny. The Mop-headed Acacia.

VISCOSA, vis-*ko*-sa; viscid or gummy.

**Rochea,** *ro*-she-a; named after French botanist, François Delaroche. Greenhouse succulents.

COCCINEA, kok-*sin*-e-a, scarlet.

FALCATA, fal-*ka*-ta, sickle-shaped—the leaves.

JASMINEA, jas-*min*-e-a, jessamine-like.

**Rodgersia,** rod-*jer*-se-a; named after Admiral Rodgers (U.S.A.), who commanded the expedition in which R. podophylla was discovered. Hardy perennials.

ÆSCULIFOLIA, es-ku-lif-*o*-le-a, horse-chestnut (Æsculus)-leaved.

PINNATA, pin-*na*-ta, pinnate—the leaves.

PODOPHYLLA, pod-o-*fil*-la, foot-stalked leaves.

SAMBUCIFOLIA, sam-bu-kif-*ol*-e-a, sambucus (Elder)-leaved.

TABULARIS, tab-ul-*ar*-is, table-like—the leaves.

**Roella,** ro-*el*-a; after G. Roelle, a Dutch botanist. Greenhouse flowering shrubs.

CILIATA, seel-e-*a*-ta, having a fringe of hairs like eyelashes.

ELEGANS, *el*-e-ganz, elegant.

**Romneya,** *rom*-ne-a; after T. Romney Robinson, an Irish astronomer. Subshrubby perennials.

COULTERI, *kole*-ter-i, after Coulter.

TRICHOCALYX, trik-*o*-ka-liks, hairy-calyxed.

**Romulea,** rom-*u*-le-a; said to be named after Romulus, the founder of Rome. Half-hardy bulbs.

BULBOCODIUM, bul-bo-*ko*-de-um, old generic name, Gr. *bolbos*, a bulb, and *kodion*, a little fleece, the hair-like covering of the bulb.

**Rondeletia,** ron-del-*ee*-te-a; after G. Rondelet, French physician and author. Warm-house flowering shrubs.

ODORATA, od-o-*ra*-ta, fragrant.

SPECIOSA, spes-e-*o*-sa, showy.

**Rosa,** *ro*-za; the ancient Latin name for the Rose, perhaps from Celtic, *rhod*, red. Experts claim the Latins pronounced the word *ros*-a—short "o". Flowering shrubs.

ACICULARIS, a-sik-ul-*ar*-is, needle-shaped—the thorns.

ALBA, *al*-ba, white—the flowers.

ALPINA, al-*pine*-a, alpine.

ARVENSIS, ar-*ven*-sis, pertaining to fields.

BANKSIÆ, *banks*-e-e, after Lady Banks. The Banksian Rose.

BRACTEATA, brak-te-*a*-ta, flowers surrounded by bracts.

CANINA, kan-*i*-na, a dog. The Dog Rose.

CAROLINA, kar-o-*li*-na, from Carolina.

CENTIFOLIA, sen- (or ken-) tif-*o*-le-a, hundred-leaved—the numerous petals. The Cabbage Rose.

CINNAMOMEA, sin-na-*mo*-me-a, cinnamon, the flowers spicily fragrant.

DAMASCENA, dam-as-*se*- (or ke-) na, damask. The Damask Rose.

ECÆ, *ek*-e, adapted from "E. C. A.," Mrs. Aitchison's initials, her husband having introduced the species.

FEROX, *fe*-roks, fierce, very prickly.

GALLICA, *gal*-ik-a, French.

HUGONIS, hu-*go*-nis, introduced by Pater Hugo (Hugh Scallan).

HUMILIS, *hum*-il-is, low-growing.

INDICA, *in*-dik-a, Indian. The Chinese or Monthly Rose.

LÆVIGATA, lev-e-*ga*-ta, smooth—the leaves.

LUCIDA, *lu*-sid-a, bright—the glossy foliage.

LUTEA, *lu*-te-a, yellow. The Austrian Briar.

MACROPHYLLA, mak-ro-*fil*-la, large-leaved.

MICROPHYLLA, mi-krof-*il*-la, small-leaved.

**Rosa** (*continued*)

MOSCHATA, mos-*ka*-ta, musk. The Musk Rose.

MOYESII, moy-*eez*-ei, after Rev. J. Moyes, a missionary in China.

MULTIFLORA, mul-tif-*lor*-a, many-flowered.

NITIDA, *nit*-id-a, glossy or shining—the leaves.

NOISETTIANA, noy-set-te-*a*-na, after M. Philippe Noisette. The Noisette Rose.

NUTKANA, nut-*ka*-na, from Nootka, N.W. America.

POLYANTHA, pol-e-*an*-tha, many-flowered.

POMIFERA, pom-*if*-er-a, apple-bearing; the fruits are very large.

RUBIGINOSA, roo-be-gin- (or jin-)*o*-sa, rusty—the foliage. The Sweet Briar.

RUBRIFOLIA, roo-bre-*fo*-le-a, red-leaved.

RUGOSA, roo-*go*-sa, wrinkled—the leaves. The Japanese Briar.

SERICEA, ser-*is*-e-a, silky—the underparts of the leaves.

SETIGERA, set-*ij*-er-a, bearing bristles.

SICULA, *sik*-ul-a, of Sicily, whence it comes.

SPINOSISSIMA, spi-no-*sis*-sim-a, most spiny. The Burnet or Scotch Rose.

TOMENTOSA, to-men-*to*-sa, downy—the leaves.

WICHURAIANA, witch-oo-ra-e-*a*-na, after Max E. Wichura, German botanist. The Rambler Rose.

**Roscoea**, ros-*ko*-e-a; after William Roscoe, founder of Liverpool Botanic Garden. Hardy perennials.

CAUTLEOIDES, kawt-le-*oy*-dez, Cautlea-like.

**Rosmarinus**, ros-mar-*e*-nus; from L. *ros*, dew (spray), and *marinus*, sea. Often inhabiting sea-cliffs.

OFFICINALIS, of-fis-in-*a*-lis, of the shop (herbal). The Rosemary.

**Rubus**, *roo*-bus; old Roman name, probably derived from L. *ruber*, red, the colour of the fruits of many species. Climbing or trailing and upright—stemmed shrubby plants; fruiting and ornamental.

BIFLORUS, bif-*lo*-rus, two-flowered.

DELICIOSUS, de-lis-e-*o*-sus, delicious—referring to the beauty of the blossoms.

FRUTICOSUS, frut-ik-*o*-sus, shrubby. The Blackberry.

GIRALDIANUS, *jer*-al-de-*a*-nus, after Giraldi, a botanist who discovered it.

IDÆUS, id-*a*-us, of Mt. Ida. The Raspberry.

I. LOGANI, lo-*gan*-e, after Judge Logan. The Loganberry.

**Rubus** (*continued*)

LACINIATUS, las-in-e-*a*-tus, leaves jagged or deeply cut. The Parsley-leaved Bramble.

NUTKANUS, nut-*ka*-nus, of Nootka, N.W. America.

ODORATUS, od-or-*a*-tus, sweet scented.

PHŒNICOLASIUS, fen-*ik*-ol-*as*-e-us, purple-haired—the stems. The Wineberry.

ROSÆFOLIUS, ro-za-*fol*-e-us, rose-leaved. The Strawberry-Raspberry.

**Rudbeckia**, rood-*bek*-e-a; named after Olaf Rudbeck, a Swedish botanist. Herbaceous perennials.

HIRTA, *hir*-ta, hairy.

LACINIATA, las-in-e-*a*-ta, cut-leaved.

MAXIMA, *maks*-im-a, greatest.

NEWMANII, *new*-man-ei, after Newman.

PINNATA, pin-*na*-ta, the leaves pinnate.

PURPUREA, pur-*pur*-e-a, purple-coloured—the flowers.

SPECIOSA, spes-e-*o*-sa, showy.

**Ruellia**, roo-*el*-le-a; after John Ruelle, botanist and physician to Francis I of France. Warm-house flowering shrub.

MACRANTHA, mak-*ranth*-a, large-flowered.

**Rumex**, *roo*-meks; old Latin name for a kind of sorrel, from L. *rumo*, to suck, from the habit of Romans sucking sorrel leaves to allay thirst. Border and aquatic perennials.

ACETOSA, as- (or ak-) et-*o*-sa, acid. The Garden Sorrel.

HYDROLAPATHUM, hid-rol-a-*path*-um, growing in water. The Water Dock.

PATIENTIA, pat-e-*en*-te-a, patience. The Herb Patience.

SCUTATUS, skew-*ta*-tus, shield-like. The French Sorrel.

**Ruscus**, *rus*-kus; said to be a corruption of *bruscus*, the old herbalists' name for Butcher's Broom; possibly from Celtic *brus*, a box; and *kelen*, holly, hence the name Box Holly. Evergreen shrubby plants.

ACULEATUS, ak-u-le-*a*-tus, prickly.

HYPOGLOSSUM, hi-po-*glos*-sum, lit. under tongue, in reference to the leaf-like bract on the underside.

**Russelia**, rus-*sel*-e-a; named after Alexander Russell, British physician and traveller. Greenhouse evergreens.

ELEGANTISSIMA, el-e-gan-*tis*-sim-a, most elegant.

JUNCEA, *junk*-e-a, rush (Juncus)-like—the stems.

SARMENTOSA, sar-men-*to*-za, twiggy, or creeping with tendrils.

**Ruta,** *roo*-ta; ancient name for Rue, origin doubtful.

GRAVEOLENS, *grav*-e-ol-enz, strong-smelling. The Rue.

**Sabbatia,** sab-*ba*-te-a; after L. Sabbati, an Italian botanist. Biennials.

CAMPESTRIS, kam-*pes*-tris, of fields.

**Saccharum,** *sak*-ka-rum; from an old Gr. word for sugar. Tropical grass.

OFFICINARUM, of-fis-e-*na*-rum, of the shops or economic. The Sugar Cane.

**Saccolabium,** sak-ko-*la*-be-um; from L. *saccus,* a bag, and *labium,* a lip, the labellum or lip is like a bag. Tropical orchids.

AMPULLACEUM, am-pul-*la*-se-um, bottle-shaped.

BLUMEI, *blu*-me-i, after Blume.

GIGANTEUM, ji-*gan*-te-um, gigantic.

**Sagina,** sa-*ge*-na; ancient name of spurrey, which was originally regarded as a species of this genus. Pearlwort. Carpeting plants.

BOYDII, *boyd*-ei, after James Boyd, of Montrose.

GLABRA, *gla*-bra, smooth—the leaves.

PILIFERA AUREA, pil-*if*-er-a *aw*-re-a, hair-bearing and golden—the foliage.

PROCUMBENS, pro-*kum*-bens, procumbent.

**Sagittaria,** sag-it-*tair*-e-a; from L. *sagitta,* an arrow, in reference to the arrowhead form of the leaves in some species. Aquatics.

MONTEVIDENSIS, mon-tev-e-*den*-sis, of Montevideo.

SAGITTIFOLIA, sag-it-tif-*o*-le-a, arrow-leaved. The Arrow-head.

VARIABILIS, var-e-*ab*-il-is, variable, many forms.

**Saintpaulia,** saint-*paw*-le-a; after Baron Walter von St. Paul, who discovered it. Warm-house perennial.

IONANTHA, i-on-*an*-tha, violet-flowered.

**Salisburia,** sal-is-*bur*-e-a; after R. A. Salisbury, an English botanist. Coniferous trees, now Ginkgo.

ADIANTIFOLIA, ad-e-an-tif-*o*-le-a, leaves like Adiantum (maidenhair fern) pinnules. The Maidenhair Tree.

**Salix,** *sa*-liks; Lat. name for a willow, possibly from Celtic, *sal,* near, and *lis,* water. Trees and shrubs.

ALBA, *al*-ba, white—the leaf.

ARBUSCULA, ar-*bus*-ku-la, a small tree or bush.

**Salix** (*continued*)

BABYLONICA, bab-e-*lon*-ik-a, of Babylon. The Weeping Willow.

Though not a native of Babylon, it is accepted that this tree was introduced to England (1730) from Asia Minor.

CAPREA, *kap*-re-a, a goat. The Goat Willow or Sallow.

CINEREA, sin-er-*e*-a, ash-coloured.

CŒRULEA, se-*ru*-le-a, blue, underparts of the leaves. The Bat Willow.

FRAGILIS, *fraj*-il-is, brittle, hence Eng. Crack Willow.

HERBACEA, her-*ba*-se-a, herbaceous—misleading name, this being a true shrub, the smallest in Britain.

INCANA, in-*ka*-na, hoary with downy hairs.

LANATA, lan-*a*-ta, woolly—leaves and young wood.

PENTANDRA, pent-*an*-dra, five-stamened.

REPENS, *re*-penz, creeping.

RETUSA, re-*tew*-sa, blunt—the leaves.

VIMINALIS, vim-in-*a*-lis, slender or twiggy. The Osier.

VITELLINA, vit-el-*le*-na, orange-yellow—the twigs.

**Salpiglossis,** sal-pi-*glos*-sis; from Gr. *salpigx,* a tube, and *glossa,* a tongue, refers to the style in the tube of the corolla. Tender annual.

SINUATA, sin-u-*a*-ta, having a deeply waved margin. The Scalloped Tube-tongue.

**Salvia,** *sal*-ve-a; from L. name used by Pliny, meaning safe, unharmed, referring to medicinal properties. Shrubs and herbaceous plants.

ARGENTEA, ar-*jen*-te-a, silvery—the foliage.

AZUREA GRANDIFLORA, az-*u*-re-a gran-dif-*lo*-ra, large-flowered and azure blue.

CANESCENS, kan-*es*-senz, hoary.

CARDUACEA, kar-du-*a*-ce-a, thistle-like.

COCCINEA, kok-*sin*-e-a, scarlet, the flowers.

DICHROA, *dik*-ro-a, two-coloured, the flowers.

FARINACEA, far-in-*a*-se-a, mealy.

GLUTINOSA, glu-tin-*o*-sa, sticky or glutinous.

GRAHAMII, *gra*-am-ei, after Graham.

HEERI, he-er-i, after Heer.

HORMINUM, hor-*mi*-num, an old generic name of doubtful origin.

INVOLUCRATA, in-vol-u-*kra*-ta, having an involucre.

OFFICINALIS, of-fis-in-*a*-lis, of the shop. The Common Sage.

PATENS, *pa*-tens, spreading.

**Salvia** (*continued*)

PITCHERI, *pitch*-er-i, after Pitcher.

RUTILANS, *root*-e-lanz, shining with ruddy gleam.

SCLAREA, *sklar*-e-a, clary, old name for sage, L. word meaning clear—its use in eye lotions.

SPLENDENS, *splen*-dens, splendid, showy.

TURKESTANICA, tur-kes-*tan*-ik-a, of Turkestan.

ULIGINOSA, u-lij-in-*o*-sa, growing in swamps.

VIRGATA, vir-*ga*-ta, with willowy twigs.

**Salvinia**, sal-*vin*-e-a; after Professor Salvini of Florence. Greenhouse aquatics.

AURICULATA, aw-rik-ul-*a*-ta, like a little ear.

NATANS, na-tanz, floating.

**Sambucus**, sam-*bu*-kus; ancient L. name of elder, said to be derived from L. *sambuca*, the name of a musical instrument which was made of Elder wood. Foliage and fruiting shrubs.

NIGRA, *ni*-gra, black—the fruits. The Elder.

RACEMOSA, ras-e-*mo*-sa, flowers in racemes.

**Samolus**, *sa*-mo-lus; name used by Pliny; several derivations suggested; possibly from Celtic *san*, health, and *mos*, a pig. alluding to its value as a food for pigs. Herbaceous perennial.

REPENS, *re*-penz, creeping.

**Sanchezia**, san-*ke*-ze-a; after Jos Sanchez. Hothouse flowering perennial.

NOBILIS, *no*-bil-is, of fine appearance.

**Sanguinaria**, san-gwin-*air*-e-a; from L. *sanguis*, blood, the sap being a red colour. Tuberous perennials.

CANADENSIS, kan-a-*den*-sis, of Canada. The Blood-root.

**Sanseviera**, san-sev-e-*e*-ra after Raimond de Sangro, Prince of Sanseviero. Greenhouse and room foliage plants. Bowstring Hemp.

CYLINDRICA, sil-*in*-drik-a, cylindrical.

GUINEENSIS, gwin-e-*en*-sis, of Guinea.

ZEYLANICA, zey-*lan*-ik-a, of Ceylon.

**Santolina**, san-tol-*e*-na; derivation doubtful. Sub-shrubs.

CHAMÆCYPARISSUS, kam-e-sip-ar-*is*-sus, an old name meaning ground-cypress. The Lavender Cotton.

INCANA, in-*ka*-na, hoary.

VIRIDIS, *vir*-id-is, green, the others being silvery.

**Sanvitalia**, san-vit-*a*-le-a; after the Italian house of Sanvitali, of Parma. Hardy annual.

PROCUMBENS, pro-*kum*-benz, procumbent.

**Saponaria**, sap-on-*air*-e-a; from L. *sapo*, soap, the bruised leaves of S. officinalis producing a lather, and once used as a soap substitute. Annuals and herbaceous perennials.

CÆSPITOSA, ses-pit-*o*-sa, tufted closely, turfy.

CALABRICA, kal-*ab*-rik-a, of Calabria.

OCYMOIDES, o-sim-*oy*-dez, Ocymum-like.

OFFICINALIS, of-fis-in-*a*-lis, of the shop. The Soapwort.

PULVINARIS, pul-vin-*ar*-is, cushiony.

VACCARIA, vak-*kar*-re-a, old generic name meaning cow-herb.

**Sarcococca**, sar-ko-*kok*-ka; from Gr. *sarx*, flesh, and *kokkos*, a berry, the fruits being fleshy. Dwarf shrubs.

RUSCIFOLIA, rus-ke-*fo*-le-a, Ruscus-leaved.

SALIGNA, sal-*ig*-na, willow-like.

**Sarmienta**, sar-me-*en*-ta; after Martius Sarmiento, a Spanish botanist. Greenhouse trailer.

REPENS, *re*-penz, creeping.

**Sarracenia**, sar-ra-*se*-ne-a; after Dr. Sarrasin, a physician of Quebec. Half-hardy perennials.

DRUMMONDII, *drum*-mond-ei, after Drummond.

FLAVA, *fla*-va, yellow.

PURPUREA, pur-*pur*-e-a, purple.

VARIOLARIS, var-e-o-*lar*-is, pimpled.

**Satureia**, sat-u-*re*-a; old Lat. for savory, possibly from the Arabic *sattar*, a name applied to labiates in general. Herbs.

HORTENSIS, hor-*ten*-sis, of the garden. The Summer Savory.

MONTANA, mon-*ta*-na, mountain. The Winter Savory.

**Satyrium**, sat-*er*-e-um, from Gr. *satyrus*, a satyr. Half-hardy terrestrial orchids.

CARNEUM, *kar*-ne-um, flesh-coloured.

CORIIFOLIUM, kor-e-if-*ol*-e-um, Corræa-like leaves.

**Sauromatum**, saw-*rom*-a-tum; from Gr. *sauros*, a lizard, in reference to the spotted spathe. Tuberous perennial.

GUTTATUM, gut-*ta*-tum, spotted—the stems. The Monarch of the East.

**Saururus,** sau-*ru*-rus; from Gr. *sauros*, a lizard, and *oura*, a tail—the form suggested by the inflorescence. Aquatics.

CERNUUS, ser (or *ker*)-nu-us, nodding or drooping—the flowers. The Lizard's Tail.

CHINENSIS, tshi-*nen*-sis, of China.

**Saussurea,** saus-*su*-re-a; after N. T. de Saussure, the Swiss philosopher. Herbaceous perennials.

ALPINA, al-*pine*-a, alpine.

PYGMÆA, *pig*-me-a, of small size.

**Saxegothea,** saks-*go*-the-a; named in honour of Prince Albert, consort of Queen Victoria, who was Prince of Saxe-Coburg and Gotha. Coniferous tree. Prince Albert's Yew.

CONSPICUA, kon-*spik*-u-a, conspicuous, or remarkable.

**Saxifraga,** sax-e-*fra*-ga (correctly, saks-*if*-ra-ga; from L. *saxum*, a rock (or stone), and *frango*, to break; application disputed. Annuals and herbaceous perennials.

AIZOIDES, ay-*zoy*-dez, aizoon-like.

AIZOON, ay-*zo*-on, ever-living, foliage not fading quickly.

ALTISSIMA, al-*tis*-sim-a, very high, tall of stature.

ANDREWSII, *an*-drews-ei, after Andrews.

ARETIOIDES, ar-et-e-*oy*-dez, Aretia-like.

ASPERA, *as*-per-a, rough—the foliage.

BIFLORA, bif-*lor*-a, two-flowered.

BURSERIANA, bur-ser-e-*a*-na, commemorating Joachim Burser, a doctor of Saxony.

CÆSIA, *se*-se-a, grey-leaved.

CÆSPITOSA, ses-pit-*o*-sa, tufted.

CERATOPHYLLA, ser-at-of-*il*-la, stag's-horn-leaved.

CERNUA, *ser*-nu-a, drooping—the flowers.

COCHLEARIS, kok-le-*ar*-is, shell-like—the foliage.

CORDIFOLIA, kor-dif-*o*-le-a, heart-shaped leaves.

CORTUSÆFOLIA, kor-tu-se-*fo*-le-a, Cortusa-leaved.

COTYLEDON, kot-e-*le*-don, old name, meaning cup-shaped (leaf), usually applied to the lobe of a seed.

CRASSIFOLIA, kras-sif-*o*-le-a, thick-leaved.

CUNEATA, ku-ne-*a*-ta, wedge-shaped—the leaves.

CUNEIFOLIA, ku-nei-*fo*-le-a, wedge-leaved.

CYMBALARIA, sim-bal-*ar*-e-a, old generic name.

DIAPENSIOIDES, di-a-pen-se-*oy*-dez, like

**Saxifraga** (*continued*)

Diapensia, a very small Lapland shrub.

FREDERICI-AUGUSTI, fred-er-e-ki-aw-*gus*-ti, Frederic Augustus, this name and the following commemorate members of the ex-Royal House of Austria.

FERDINANDI-COBURGI, fer-din-*an*-di-*ko*-bur-gi, Ferdinand of Coburg.

FLORULENTA, flor-u-*len*-ta, full of flower.

FORTUNEI, *for*-tune-i, after Fortune, plant collector.

GERANIOIDES, jer-a-ne-*oy*-dez, geranium-like.

GEUM, *je*-um, obsolete name, signifying to stimulate (ancient medicine).

GLOBULIFERA, glob-ul-*if*-er-a, globe-bearing, the gem-buds being a distinctive feature.

GRANULATA, gran-u-*la*-ta, granulated—the small grain-like root-tubers.

GRISEBACHII, grise-*bach*-ei, after Grisebach, a professor of botany, Göttingen.

HOSTII, *host*-ei, after Host, a physician of Austria.

HYPNOIDES, hip-*noy*-dez, moss-like.

IMBRICATA, im-bre-*ka*-ta, the dense, overlapping foliage.

INCRUSTATA, in-krus-*ta*-ta, encrusted, *i.e.*, with silvery scales.

JUNIPERIFOLIA, jew-nip-er-if-*o*-le-a, juniper-like leaves.

LÆVIS, *le*-vis, smooth, the leaves.

LIGULATA, lig-ul-*a*-ta, strap-leaved.

LINGULATA, ling-ul-*a*-ta, tongue-leaved.

LONGIFOLIA, long-if-*o*-le-a, long-leaved.

MARGINATA, mar-jin-*a*-ta, margined—white leaf edges.

MAWEANA, maw-e-*a*-na, after Geo. Maw, plant collector.

MEDIA, *me*-de-a, medium, in allusion to the altitude of the plant's natural range, or its stature.

MOSCHATA, mos-*ka*-ta, musky.

MUSCOIDES, mus-*koy*-dez, moss-like.

NIVALIS, niv-*a*-lis, of snowy regions.

OPPOSITIFOLIA, op-pos-it-if-*o*-le-a, opposite-leaved.

PEDEMONTANA, ped-e-mon-*ta*-na, from Piedmont.

PELTATA, pel-*ta*-ta, shield-shaped—the leaves.

PURPURASCENS, pur-pur-*as*-senz, purple—the flowers.

ROTUNDIFOLIA, ro-tun-dif-*o*-le-a, round-leaved.

SANCTA, *sang*-ta, sacred, holy; application obscure.

SARMENTOSA, sar-men-*to*-za, with long, slender runners.

SIBTHORPII, sib-*thorp*-ei, Sibthorp, professor of botany.

**Saxifraga** (*continued*)

SPATHULATA, spath-ul-*a*-ta, spathulate—spoon-shaped leaves.

SQUARROSA, skwa-*ro*-za, scurfy—the foliage.

STELLARIS, stel-*lar*-is, starry—the flowers.

STRACHEYI, *stray*-ke-i, after Strachey.

TAYGETEA, tay-*ge*-tc-a, from Taygetus (ancient geography).

TELLIMOIDES, tel-li-*moy*-dez, Tellima-like.

TENELLA, ten-*el*-la, delicate, fine in texture.

TOMBEANENSIS, tom-be-an-*en*-sis, of Tombea, Italy.

TRICUSPIDATA, trik-us-pe-*da*-ta, three-toothed—the leaf lobes.

UMBROSA, um-*bro*-za, shade-loving.

VALDENSIS, val-*den*-sis, from Mt. Balde, N. Italy.

VALENTINA, val-en-*te*-na, from Valentia.

VANDELLII, van-*del*-ei, after Vandelli, Portuguese botanist.

**Scabiosa**, ska-be-*o*-sa; from L. *scabies*, itch, the plant once being regarded as a remedy for skin diseases. Annual and perennial herbs. Pincushion Flower.

ATROPURPUREA, atro-pur-*pur*-e-a, deep purple.

CAUCASICA, kaw-*kas*-ik-a, Caucasian.

COLUMBARIA, kol-um-*bar*-e-a, old name meaning dove-coloured.

OCHROLEUCA, *ok*-ro-*loo*-ka, yellowish white.

PTEROCEPHALA, ter-o-*sef*-al-a, old generic name signifying a winged head, alluding to the form of the flower.

**Schivereckia**, shiv-er-*ek*-e-a; after A. Schivreck, a Russian botanist. Rock plants.

PODOLICA, pod-*ol*-ik-a, of Podolia, S.W. Russia.

**Schizandra**, ski-*zan*-dra; from Gr. *schizo*, to cut, and *aner*, a man, in reference to the split stamens. Sometimes spelt Schisandra. Twining shrubs.

CHINENSIS, tshi-*nen*-sis, Chinese.

COCCINEA, kok-*sin*-e-a, scarlet—the flowers.

HENRYI, *hen*-re-e, after Dr. Augustine Henry.

**Schizanthus**, skiz-*an*-thus; from Gr. *schizo*, to cut, and *anthos*, a flower, the petals deeply fringed. Greenhouse annuals.

PINNATUS, pin-*na*-tus, foliage pinnate, the best-known species.

RETUSUS, re-*tu*-sus, blunted, or notched, the petals.

WISETONENSIS, wis-ton-*en*-sis, of Wiseton.

**Schizocodon**, skiz-ok-*o*-don; from Gr. *schizo*, to cut, and *codon*, a bell, the bell-shaped flowers being deeply cut or fringed. Rock or woodland plant.

SOLDANELLOIDES, sol-dan-el-*loy*-dez, Soldanella-like.

**Schizopetalon**, skiz-o-*pet*-a-lon; from Gr. *schizo*, to cut, and *petalon*, a petal, the fringed flowers. Annual.

WALKERI, *waw*-ker-i, after Walker.

**Schizophragma**, skiz-o-*frag*-ma; from Gr. *schizo*, to cut, and *phragma*, wall of an enclosure, in reference to the curious splitting of the seed capsules. Shrubs.

HYDRANGEOIDES, hy-dran-je-*oy*-dez, hydrangea-like.

**Schizostylis**, skiz-*os*-til-is; from Gr. *schizo*, to cut, and *stylos*, a style, the latter being deeply divided. Half-hardy perennial.

COCCINEA, kok-*sin*-e-a, scarlet.

**Schlumbergera**, sklum-*ber*-ger-a; after Frederick Schlumberger, a student of plants. Greenhouse cacti.

GÆRTNERI, *gert*-ner-i, after Gærtner.

RUSSELLIANUM, rus-sel-le-*a*-num, after Russell.

**Schubertia**, shu-*ber*-te-a; after H. von Schubert, a botanist. Warm-house flowering twiners.

GRANDIFLORA, gran-dif-*lo*-ra, large-flowered.

**Sciadopitys**, si-a-*dop*-it-is; from Gr. *skias*, a parasol or sunshade, and *pitus*, a fir tree, the whorled leaves like the ribs of an umbrella.

VERTICILLATA, ver-tis-il-*la*-ta, whorled—the leaves. The Umbrella Pine.

**Scilla**, *sil*-la; ancient Gr. or L. name. Hardy and tender bulbs.

AMŒNA, am-*e*-na, pleasing.

AUTUMNALIS, aw-tum-*na*-lis, of autumn.

BIFOLIA, bif-*o*-le-a, two-leaved.

CAMPANULATA, kam-pan-u-*la*-ta, bell-flowered.

FESTALIS, fes-*ta*-lis, gay.

HISPANICA, his-*pan*-ik-a, of Spain. The Spanish Bluebell.

HYACINTHOIDES, hy-a-sinth-*oy*-dez, hyacinth-like.

ITALICA, it-*al*-ik-a, Italian.

LEUCOPHYLLA, lew-kof-*il*-a, white-leaved.

LIGULATA, lig-ul-*a*-ta, tongue-shaped.

PATULA, *pat*-u-la, spreading.

PERUVIANA, per-u-ve-*a*-na, Peruvian, misleading, for this species is from the Mediterranean region.

PRATENSIS, pra-*ten*-sis, of meadows.

**Scilla** (*continued*)

PUSCHKINIOIDES, pus-kin-e-*oy*-dez, Puschkinia-like.

SIBIRICA, si-*bir*-ik-a, of Siberia.

VERNA, *ver*-na, spring.

**Scirpus**, *sir*-pus; or *sker*-pus; the old Latin name for a rush or reed. Bog plants.

CERNUUS, *ser*-nu-us, drooping.

LACUSTRIS, lak-*us*-tris, of lakes.

SETACEOUS, se-*ta*-se-us, bristly.

TABERNÆMONTANI ZEBRINUS, tab-*er*-ne-mon-*ta*-ne-ze-*bry*-nus, after Tabernemontanus and striped like a zebra.

TRIQUETER, tre-*kwe*-ter, stems three-angled.

**Sclerocactus**, skler-o-*kak*-tus; possibly from Gr. *scleros*, hard (cruel), and *cactus* —the spines, being hooked. Greenhouse cacti.

POLYANCISTRUS, pol-e-an-*sis*-trus, many angles or hooks.

**Scolopendrium**, skol-o-*pen*-dre-um; Gr. *scolopendra*, a centipede, a name originally applied to the Ceterach fern, which resembled that creature. The ripe sori are very suggestive of centipedes. Hardy evergreen fern.

VULGARE, vul-*gar*-e, common. The Harts-tongue Fern.

V. CRISPUM, kris-pum, curled or wavy— the frond margins. The most popular of very numerous varieties.

**Scolymus**, *skol*-im-us; old Gr. name used by Hesiod, possibly from Gr. *skolos*, a thorn, these plants being spiny. Herbaceous perennials.

HISPANICUS, his-*pan*-ik-us, of Spain.

MACULATUS, mak-ul-*a*-tus, spotted.

**Scorzonera**, skor-zon-*e*-ra; from old Fr. *scorzon*, serpent, the plant once being regarded as a remedy for snake-bite. Edible-rooted perennial.

HISPANICA, his-*pan*-ik-a, of Spain. The Scorzonera.

**Scutellaria**, sku-tel-*lar*-e-a; from L. *scutella*, a dish, referring to the form of the persistent calyx. Greenhouse and hardy perennials. The Skull-caps.

ALPINA, al-*pine*-a, alpine.

BAICALENSIS, bi-kal-*en*-sis, from Baikal, Russia.

COCCINEA, kok-*sin*-e-a, scarlet.

GALERICULATA, gal-er-ik-ul-*a*-ta, small capped.

INDICA JAPONICA, *in*-dik-a jap-*on*-ik-a, Indian (Japanese form).

**Seaforthia,** see-*forth*-e-a; after Francis, Lord Seaforth, a patron of botany. Greenhouse palm.

ELEGANS, *el*-e-ganz, elegant.

**Sedum,** *se*-dum; from L. to assuage, from the healing properties of the house-leek to which the name was applied, as well as to the stonecrop, by Roman writers. Succulent greenhouse and hardy evergreen and deciduous rock plants.

ACRE, *ak*-re, biting or sharp to the taste. The Common Stonecrop.

AIZOON, ay-*zo*-on, ever living.

ALBUM, *al*-bum, white. The White Stonecrop.

ALTISSIMUM, al-tis-*se*-mum, very tall.

AMPLEXICAULE, am-pleks-e-*kaw*-le, stem-clasping—the leaves.

ANACAMPSEROS, an-a-*kamp*-ser-os, an old generic name, meaning "to cause love to return" (Greek mythology).

ANGLICUM, *ang*-lik-um, of England.

ANOPETALUM, an-o-*pet*-a-lum, upward-growing—the petals.

BREVIFOLIUM, brev-if-*o*-le-um, short-leaved.

CÆRULEUM, ser-*u*-le-um (or ker-*u*-le-um), blue.

CARNEUM VARIEGATUM, kar-*ne*-um var-e-eg-*a*-tum, flesh tinted and variegated.

COMPACTUM, kom-*pak*-tum, compact— the habit of growth.

DASYPHYLLUM, das-e-*fil*-lum, thick-leaved.

DENDROIDEUM, den-dro-*id*-e-um, tree-like.

DIVERGENS, di-*ver*-jenz, spreading.

ELLACOMBIANUM, el-la-ko-me-*a*-num, after Canon Ellacombe.

EWERSII, ew-*ers*-ei, after Ewers, botanist and traveller.

FARINOSUM, far-in-*o*-sum, covered with farina, or meal.

GLAUCUM, *glaw*-kum, sea-blue or glaucous.

HISPANICUM, his-*pan*-ik-um, of Spain.

HUMIFUSUM, hum-if-*ew*-sum, spread over the ground.

KAMTSCHATICUM, kamts-*kat*-ik-um, of Kamchatka.

LIEBMANNIANUM, leeb-man-e-*a*-num, after F. Liebmann, Dutch botanist and collector.

LINEARE VARIEGATUM, lin-e-*ar*-e var-e-eg-*a*-tum, narrow leaved and variegated.

LYDIUM, lid-*e*-um, old name, possibly alluding to Lydia, Asia Minor.

MAXIMUM, *maks*-e-mum, largest.

MAGELLENSE, maj-el-*en*-se, of Monte Majella, Italy.

**Sedum** (*continued*)

MIDDENDORFFIANUM, mid-den-dor-fe-*a*-num, after A. T. von Middendorff, introducer of many Siberian plants.

MORANENSE, mor-an-*en*-se, from Real de Moran, Mexico.

MULTICEPS, *mul*-te-seps, many-headed.

NEVII, *ne*-vei, after the Rev. Dr. Nevius, its discoverer.

OBTUSATUM, ob-tew-*sa*-tum, obtuse or blunt—the leaves.

OREGANUM, or-e-*ga*-num, Oregon, first found by River Oregon.

PALMERI, *pah*-mer-i, after Dr. Palmer, Mexican botanical explorer.

PILOSUM, pil-*o*-sum, hairy—the leaves.

POPULIFOLIUM, pop-u-lif-*o*-leum, Populus (poplar)-leaved.

PRÆLATUM, pre-*al*-tum, very high—the growth.

PRAEGERIANUM, pra-ger-e-*a*-num, after Dr. Lloyd Prager.

PRIMULOIDES, prim-u-*loy*-dez, primula-like.

PULCHELLUM, pul-*kel*-lum, beautiful, pleasing (dim.).

REFLEXUM, re-*fleks*-um, leaves bent back, recurved.

RHODIOLA, *rod*-e-o-la; from L. *rhodia radix*, the rosy odour and tint of the root stocks. The Rose-root.

ROSEUM, *ro*-ze-um, the fragrance of the fleshy root-stock, hence Rose-root (Rhodiola).

RUPESTRE, roo-*pes*-tre, rock-breaking.

SEMPERVIVOIDES, sem-per-viv-*oy*-dez, Sempervivum-like.

SEXANGULARE, seks-ang-ul-*ar*-e, the leaves in six rows.

SIEBOLDII, se-*bold*-ei, after P. F. von Siebold.

SPECTABILE, spek-*tab*-il-e, showy, striking.

SPURIUM, *spew*-re-um, false or doubtful, possibly in reference to its many false names.

STAHLII, *stah*-lei, after Professor E. Stahl of Jena.

STOLONIFERUM, sto-lon-*if*-er-um, runner-bearing—the creeping stems.

TELEPHIUM, te-*lef*-e-um, old name derived from Telephus, son of Hercules.

TERNATUM, ter-*na*-tum, leaves in threes.


**Selaginella**, sel-a-jin-(or gin-)*el*-la; diminutive of Selago, ancient name of a lycopodium. Moss-like branching herb.

APUS, *a*-pus, stalkless.

CÆSIA, *se*-se-a, grey.

CUSPIDATA, kus-pid-*a*-ta, short-pointed.

DENTICULATA, den-tik-ul-*a*-ta, finely toothed.


**Selaginella** (*continued*)

GRANDIS, *gran*-dis, grand.

KRAUSSIANA, kraus-se-*a*-na, after Krauss.

K. AUREA, *aw*-re-a, golden.

K. VARIEGATA, var-e-eg-*a*-ta, variegated (with white).

LEPIDOPHYLLA, lep-id-*of*-il-la, slender leaved. The Resurrection Plant.

MARTENSII, mar-*ten*-sei, after Martens.

UNCINATA, un-sin-*a*-ta, hooked—end of leaves.

WILLDENOVII, wil-den-*o*-vei, after Willdenow.


**Selenicereus**, sel-en-e-*se*-re-us; from Gr. *selene*, the moon, and *cereus*—the flowers opening at night. Greenhouse climbing cacti.

CONIFLORUS, ko-nif-*lor*-us, cone-like flower buds.

GRANDIFLORUS, gran-dif-*lor*-us, large-flowered. The Queen of the Night.

MACDONDALDIÆ, mak-don-ald-e-e, after Mrs. Macdonald.


**Selenipedium**, sel-en-e-*ped*-e-um; from Gr. *selenis*, a little crescent, and *podion*, a slipper, the shape of the labellum. Greenhouse terrestrial orchids.

CAUDATUM, kaw-*da*-tum, tailed.

DOMINIANUM, dom-in-e-*a*-num, after Dominy.

SEDENII, sed-*e*-nei, after Mr. John Seden, an early orchid hybridizer.


**Sempervivum**, sem-per-*vi*-vum; from L. *semper*, always, and *vivo*, alive, alluding to the tenacity of life common to these plants. Greenhouse and hardy succulents.

ARACHNOIDEUM, ar-ak-*noy*-de-um, cobwebbed.

ARBOREUM, ar-*bor*-e-um, tree-like.

ARENARIUM, ar-en-*a*-re-um, sand-loving.

CÆSPITOSUM SPATHULATUM, ses-pit-*o*-sum spath-ul-*a*-tum, tufted and with spoon-shaped leaves.

CALCARATUM, kal-kar-*a*-tum, spurred.

CANARIENSIS, kan-ar-e-*en*-sis, of the Canary Islands.

COMOLLII, kom-*ol*-lei, after Comoll.

DOMESTICUM FOLIIS VARIEGATIS, do-*mes*-tik-um fol-*e*-is var-e-eg-*a*-tis, variegated leaves and found in dwelling houses.

FUNCKII, funk-ei; after Funck, an apothecary.

HAWORTH, ha-*worth*-e, after Haworth.

HEUFFELII, *huf*-fel-ei, after Heuffeli, Hungarian botanist.

HOLOCHRYSUM, hol-o-*kry*-sum, wholly golden.

**Sempervivum** (*continued*)

MONTANUM, mon-*ta*-num, mountain.

RUTHENICUM, ru-*then*-ik-um, Russian.

SOBOLIFERUM, sob-ol-*if*-er-um, stolon-bearing.

TABULAEFORME, tab-ul-e-*for*-me, table-shaped.

TECTORUM, tek-*tor*-um, of roofs. The Houseleek.

T. CALCAREUM, kal-*kar*-e-um, chalk-loving.

T. GLAUCUM, *glaw*-kum, sea-green.

**Senecio**, sen-*e*-se-o (or sen-*e*-she-o); from L. *senex*, old (an old man), in allusion to the grey and hoary seed pappus. Annuals, hardy and greenhouse perennials, and shrubs. Most florists' cinerarias (now in this genus) commonly grown are hybrids, but some of the species from which these were raised are given below.

ABROTANIFOLIUS, ab-*rot*-an-if-*o*-le-us, Artemisia (abrotanum)-leaved.

ADONIDIFOLIUS, ad-on-id-if-*o*-le-us, Adonis-leaved.

CINERARIA, sin-er-*ar*-e-a, old name, signifying ashen.

CLIVORUM, kli-*vor*-um, of the hills.

CRUENTUS, kru-*en*-tus, blood-red—the flowers. Parent of the florists' Cineraria.

DORONICUM, dor-*on*-ik-um, old generic name for this or similar plant.

ELEGANS, *el*-e-ganz, elegant.

GREYI, gra-i, after Grey.

INCANUS, in-*ka*-nus, hoary or grey.

KÆMPFERI, kem-fer-i, of Kæmpfer, a German botanist.

LANATUS, lan-*a*-tus, woolly.

LAXIFOLIUS, laks-e-*fo*-le-us, loosely spread leaves.

MACROGLOSSUS, mak-ro-*glos*-sus, large tongued. The Cape Ivy.

MARITIMA, mar-*it*-im-a, pertaining to the sea, maritime.

PULCHER, *pul*-ker, beautiful.

TANGUTICUS, tan-*gu*-tik-us, Tangusian, Siberia.

VEITCHIANUS, veech-e-*a*-nus, after Veitch.

WILSONIANUS, wil-son-e-*a*-nus, after Wilson.

**Sequoia**, se-*kwoy*-a; named after Sequoiah, inventor of the Cherokee alphabet. The Redwood. Coniferous trees.

GIGANTEA, ji-*gan*-te-a, gigantic. The Giant Tree of California.

SEMPERVIRENS, sem-per-*ver*-ens, always green.

**Serapias**, ser-*a*-pe-as; the name of an Egyptian deity and used by Pliny for this plant or an ally. Hardy terrestrial orchids.

CORDIGERA, kor-*dig*-er-a, heart-bearing.

LINGUA, *lin*-gwa, a tongue.

LONGIPETALA, long-ip-*et*-a-la, long petalled.

**Serratula**, ser-*ra*-tu-la; from L. *serrula*, a saw (lit. a little saw), in allusion to the toothed leaf-margins. Hardy herbaceous perennials.

CORONATA, kor-on-*a*-ta, crowned, the tufted flower-heads.

**Sericographis**, ser-ik-o-*graf*-is; from Gr. *serikos*, silk, and *grapho*, to write. Greenhouse flowering shrubs.

GHIESBREGHTIANA, gies-bret-e-*a*-na, after Ghiesbreght.

**Shepherdia**, shep-*her*-de-a; after J. Shepherd, curator of Liverpool Botanic Gardens. Shrub.

ARGENTEA, ar-*jen*-te-a, silvery—the leaves.

**Shortia**, *shor*-te-a; after a Dr. Short, botanist of Kentucky. Woodland and rock-garden plants.

GALACIFOLIA, ga-las-if-*o*-le-a, Galax-leaved.

UNIFLORA, u-nif-*lor*-a, one-flowered.

**Sibthorpia**, sib-*thor*-pe-a; after Professor Sibthorp, of Oxford, a distinguished botanist. Trailing plants.

EUROPÆA, u-*ro*-pe-a, European,

**Sidalcea**, sid-*al*-se-a; compound of Sida and Alcea, related genera. Herbaceous perennials.

CANDIDA, *kan*-did-a, white.

LISTERI, *lis*-ter-i, after Dr. Lister, a naturalist.

MALVÆFLORA, mal-ve-*flor*-a, mallow-(Malva)-flowered.

**Silene**, si-*le*-ne; probably from Gr. *sialon*, saliva, the gummy exudations on the stems which ward off insects. Annuals and herbaceous perennials. Catchfly.

ACAULIS, a-*kaw*-lis, stemless.

ALPESTRIS, al-*pes*-tris, alpine.

ARMERIA, ar-*meer*-e-a, old generic name meaning "near the sea."

COMPACTA, kom-*pak*-ta, compact—the flower heads.

ELIZABETHÆ, el-iz-a-*be*-the, after Elizabeth.

HOOKERI, *hook*-er-e, after W. J. Hooker, Director of Kew Gardens.

**Silene** (*continued*)
LACINIATA, las-in-e-*a*-ta, the petals cut or fringed.
MARITIMA, mar-*it*-im-a, maritime.
PENDULA, *pen*-du-la, drooping, the flowers.
QUADRIFIDA, kwod-*rif*-id-a, petals four-notched.
SCHAFTA, *shaf*-ta, origin doubtful, probably name of place in Russia where found.

**Silphium**, *sil*-fe-um; ancient name referring to the resinous juice. Herbaceous perennials.
LACINIATUM, la-sin-e-*a*-tum, cut into narrow fringe-like segments. The Compass Plant.
PERFOLIATUM, per-fo-le-*a*-tum, the leaves perfoliate or stem clasping.

**Silybum**, *sil*-ib-um; name applied by Dioscorides to some thistle-like plants. Biennial.
MARIANUM, mar-e-*a*-num, St. Mary's.

**Sinningia**, sin-*ning*-e-a; after William Sinning, gardener to the Bonn University. Greenhouse tuberous perennials.
SPECIOSA, spes-e-*o*-sa, showy. From this plant the hybrid Gloxinia has been developed.

**Sinofranchetia**, si-no-fran-*shet*-e-a; from *sino*, China, and Franchet, a French botanist and authority on Chinese plants. Climbing shrub.
SINENSIS, si-*nen*-sis, of China.

**Sinowilsonia**, si-no-wil-*so*-ne-a; from *sino*, China, and Wilson, the well-known plant collector. Shrub.
HENRYI, *hen*-re-i, after Henry, the Chinese traveller.

**Sisyrinchium**, sis-e-*rink*-e-um; old Gr. name first applied to some other plant. Small herbaceous perennials with grass-like tufted leaves. Blue-eyed Grass.
ANGUSTIFOLIUM, an-gus-tif-*o*-le-um, narrow-leaved.
BERMUDIANUM, ber-mu-de-*a*-num, of Bermuda.
FILIFOLIUM, fil-if-*o*-le-um, thread-like foliage.
GRAMINIFOLIUM, gram-in-if-*o*-le-um, grass-leaved.
GRANDIFLORUM, gran-dif-*lo*-rum, large-flowered.
STRIATUM, stri-*a*-tum, the leaves channelled or grooved.

**Sium**, *see*-um; old Gr. name for a marsh plant. The root vegetable, Skirret.
SISARUM, *sis*-ar-um, from the Arabic *dgizer*, a carrot.—The shape of the roots.

**Skimmia**, *skim*-e-a; from *skimmi*, a Japanese name for S. japonica. Berry-bearing shrubs.
FORTUNEI, *for*-tune-i, after Robert Fortune, plant collector.
JAPONICA, jap-*on*-ik-a, of Japan.

**Smilacina**, smi-las-*e*-na; diminutive of Smilax (*q.v.*), literally a little smilax, which it resembles. Herbaceous perennials.
RACEMOSA, ras-em-*o*-sa, the flowers in racemes.
STELLATA, stel-*la*-ta, star-flowered.

**Smilax**, *smi*-laks; ancient Gr. name of obscure meaning. The smilax of florists is Asparagus asparagoides. Tendril-climbing vines. The Greenbrier.
ARGYREA, ar-gy-*re*-a, silvery.
ASPERA, *as*-per-a, rough.
OFFICINALIS, of-fis-in-*a*-lis, of the shops. The Sarsparilla.

**Sobralia**, sob-*ra*-le-a; after Don F. M. Sobral, a Spanish botanist. Warm-house terrestrial orchids.
MACRANTHA, mak-*ran*-tha, large-flowered.

**Solandra**, so-*lan*-dra; after Daniel Charles Solander, LL.D., F.R.S., a Swede and companion of Sir Joseph Banks during his voyage round the world. Tropical flowing shrubs.
GRANDIFLORA, gran-dif-*lor*-a, large-flowered.

**Solanum**, so-*la*-num; name given by Pliny, the Roman naturalist, to one of the nightshades; possibly derived from L. *solamen*, a solace, from its medicinal virtues. Hardy and greenhouse annuals, herbaceous perennials, tuberous-rooted vegetable and shrubs.
CAPENSE, ka-*pen*-se, of the Cape—Cape Colony.
CAPSICASTRUM, kap-sik-*as*-trum, old generic name.
CRISPUM, *kris*-pum, curled—application obscure.
JASMINOIDES, jas-min-*oy*-dez, jasmine-like.
LOBELII, lo-*bel*-ei, of Lobel, a Belgian botanist.

**Solanum** (*continued*)

MARGINATUM, mar-jin-*a*-tum, edged—with white.

MELONGENA, mel-*on*-je-na, old name referring to the large fruits of the Egg-plant. The Egg-plant.

PSEUDO-CAPSICUM, *sue*-do-*kap*-sik-um, false capsicum.

SISYMBRIFOLIUM, sis-im-bre-*fo*-le-um, Sisymbrium-leaved.

TUBEROSUM, tu-ber-*o*-sum, tuber bearing. The Potato.

WENDLANDII, wend-*land*-ei, after Wendland.

**Soldanella**, sol-dan-*el*-la; said to be from *soldo*, an Italian coin, in allusion to the roundness of the leaf. Alpine.

ALPINA, al-*pine*-a, alpine.

MINIMA, *min*-im-a, least.

MONTANA, mon-*ta*-na, inhabiting mountains.

**Solidago**, sol-id-*a*-go; from L. *solido*, to make whole or to heal, in reference to supposed healing properties. Herbaceous perennials.

BRACHYSTACHYS, brak-e-*stak*-is, short spikes.

CANADENSIS, kan-a-*den*-sis, of Canada.

MISSOURIENSIS, mis-sour-e-*en*-sis, of Missouri, U.S.A.

VIRGA-AUREA, *vir*-ga-*aw*-re-a, old name, signifying a golden twig, hence the English name Golden-rod.

**Sollya**, *sol*-e-a; after R. H. Solly, a naturalist. Tender twining shrubs.

HETEROPHYLLA, het-er-o-*fil*-la, variously shaped leaves.

PARVIFLORA, par-vif-*lo*-ra, small-flowered.

**Sonerila**, son-er-*il*-a; from *Soneri-ila*, native Indian name. Tropical perennials, flowering and ornamental foliage.

ARGENTEA, ar-*gen*-te-a, silvery.

MACULATA, mak-ul-*a*-ta, spotted.

MARGARITACEA, mar-gar-it-*a*-ce-a, pearl-spotted.

SPECIOSA, spes-e-*o*-sa, showy.

**Sophora**, *sof*-or-a; from *sophera*, an Arabic name for a tree with pea-shaped flowers, Shrubs or trees.

JAPONICA, jap-*on*-ik-a, of Japan.

TETRAPTERA, tet-*rap*-ter-a, four-winged—the seed pod.

VICIFOLIA, vis-if-*o*-le-a, vetch (Vicia)-leaved.

**Sophronitis**, sof-ron-*i*-tis; from *sophrona*, modest, referring to the miniature

**Sophronitis** (*continued*)

cattleya-like plants and flowers. Greenhouse orchid.

GRANDIFLORA, gran-dif-*lo*-ra, grand or showy—the flowers.

**Sparaxis**, spar-*aks*-sis; from Gr. *sparasso* to tear, in reference to the lacerated spathes. Half-hardy bulbs, now mostly under "Dierama" and "Ixia."

GRANDIFLORA, gran-dif-lo-ra, fine flowered.

PULCHERRIMA, pul-*ker*-rim-a, very pretty.

TRICOLOR, *trik*-o-lor, three-coloured.

**Sparmannia**, spar-*man*-ne-a; after Andreas Sparmann, a Swedish naturalist, who accompanied Captain Cook on his second voyage round the world. Greenhouse flowering shrub.

AFRICANA, af-rik-*a*-na, of Africa.

**Spartium**, *spar*-te-um; from Gr. *spartos*, the ancient name of the plant. Flowering shrub.

JUNCEUM, *jun*-ke-um, a rush, the form of the twigs. The Spanish Broom.

**Specularia**, spek-ul-*air*-e-a; derived from L. *speculum*, a mirror. Venus's Looking-glass. Annual.

SPECULUM, spek-ul-um, a mirror. Venus's Looking-glass.

**Spergula**, *sper*-gul-a; from L. *spargo*, to scatter, alluding to the scattering of the seeds. Dwarf carpeting plant.

PILIFERA AUREA, pil-*if*-er-a *aw*-re-a, hair bearing and golden leaved.

**Sphæralcea**, sfer-*al*-se-a; from Gr. *sphaira*, a globe, and *alcea*, a mallow, alluding to the rounded form of the carpels or seed pods. Hardy perennial.

MUNROANA, mun-ro-*a*-na, after Munro.

**Sphenogyne**, sfen-*og*-in-e; from Gr. *sphen*, a wedge, and *gyne*, a woman, in allusion to the shape of the pistil. Half-hardy annuals.

SPECIOSA, spes-e-*o*-sa, showy.

**Spigelia**, spi-*je*-le-a; after Adrian van der Spiegel, a physician. Herbaceous perennials.

MARYLANDICA, mair-e-*lan*-dik-a, of Maryland, U.S.A.

**Spinacia**, spe-*na*-se-a; from L. *spina*, a prickle, in allusion to the prickly seeds. Culinary vegetable.

OLERACEA GLABRA, ol-er-*a*-se-a *glab*-ra, culinary or pot herb; smooth. The

**Spinacia** (*continued*)
Round-seeded or Summer Spinach.
O. SPINOSA, spe-*no*-sa, prickly. The Prick-ly-seeded or Winter Spinach.

**Spiræa**, spi-*re*-a; probably from Gr. *speira*, a wreath. Herbaceous perennials and shrubs.
AITCHISONII, aitch-is-*o*-nei; after Dr. Aitchison, who discovered it.
ARGUTA, ar-gu-ta, sharp.
ARUNCUS, ar-*un*-kus, old generic name of Goat's Beard.
ASTILBOIDES, as-til-*boy*-dez, astilbe-like.
BELLA, *bel*-la, pretty.
BRACTEATA, brak-te-*a*-ta, leafy bracts on flower stalks.
BULLATA, bul-*la*-ta, the leaves bullate or blistered.
BUMALDA, bu-*mald*-a, after J. A. Bumalda.
CAMSCHATICA, kams-*kat*-ik-a, of Kam-chatka.
CANA, *ka*-na, hoary.
CANESCENS, kan-*es*-senz, greyish or slightly hoary.
CANTONIENSIS, kan-ton-e-*en*-sis, of Can-ton.
CORYMBOSA, kor-imb-*o*-sa, flowers in corymbs.
CRENATA, kre-*na*-ta, crenate, or scalloped —the leaves.
DECUMBENS, de-*kum*-benz, prostrate.
DISCOLOR, *dis*-kol-or, two-coloured.
FILIPENDULA, fil-ip-*en*-du-la, thread-hanging. Presumably referring to the fibres attached to the tuberous roots.
GIGANTEA, ji-*gan*-te-a, gigantic.
JAPONICA, jap-*on*-ik-a, of Japan.
LÆVIGATA, lev-*ga*-ta, smooth-leaved.
LINDLEYANA, lind-le-*a*-na, after Prof. Lindley, a botanist.
LOBATA, la-*ba*-ta, the leaves lobed.
MEDIA, *me*-de-a, intermediate, the sta-ture. .
MENZIESII, men-zees-ei; after Menzies.
MOLLIFOLIA, mol-le-*fo*-le-a, soft-leaved.
PALMATA, pal-*ma*-ta, palmate or hand-leaved.
PRUNIFOLIA, pru-nif-*o*-le-a, plum-leaved.
REEVESIANA, reevs-e-*a*-na, after Reeves.
SALICIFOLIA, sàl-is-if-*o*-le-a, willow (Salix)-leaved.
SORBIFOLIA, sor-bif-*o*-le-a, leaved like Sorbus.
THUNBERGII, thun-*berg*-ei, after Thun-berg.
TRILOBATA, tril-o-*ba*-ta, three-lobed—the leaves.
ULMARIA, ul-*mair*-e-a, old generic name for common Meadowsweet, meaning Elm-leaved.
VAN HOUTTEI, van-*hout*-te-i, after Van Houtte.

**Sprekelia**, sprek-*e*-le-a; after J. H. von Sprekelsen, who sent the plant to Lin-næus. Greenhouse bulb.
FORMOSISSIMA, for-mos-*is*-sim-a, most beautiful.

**Stachys**, *stak*-is; from Gr. *stachus*, a spike, alluding to the pointed in-florescences. Herbaceous perennials and tuberous vegetable.
CORSICA, *kor*-sik-a, Corsican.
GRANDIFLORA, gran-dif-*lo*-ra, fine flow-ered.
LANATA, lan-*a*-ta, woolly—the foliage. The Lamb's Ear.
LAVANDULÆFOLIA, lav-an-dew-le-*fo*-le-a, lavender (Lavandula)-leaved.
TUBERIFERA, tu-ber-*if*-er-a, tuber-bearing. The Chinese Artichoke.

**Stanhopea**, stan-*ho*-pe-a; after Earl Stanhope, president of the Medico-botanical Society. Warm-house orchids.
BUCEPHALUS, bu-*kef*-al-us, bull-headed.
GRAVEOLENS, *grav*-e-ol-enz, strong smell-ing.
OCULATA, ok-ul-*a*-ta, eyed.
TIGRINA, tig-*re*-na, tiger marked.

**Stapelia**, sta-*pel*-e-a; after J. B. van Stapel, a Dutch physician. Greenhouse succulent flowering plants.
ASTERIAS, as-*teer*-e-as, star-fish-like.
BUFONIA, bu-*fo*-ne-a; toad-like.
GIGANTEA, ji-*gan*-te-a, gigantic.
GRANDIFLORA, gran-dif-*lo*-ra, large flow-ered.
HIRSUTA, hir-*su*-ta, hairy.
VARIEGATA, var-e-eg-*a*-ta, variegated—the flowers.

**Staphylea**, staf-e-*le*-a; from Gr. for a bunch of grapes, referring to the in-florescence. Shrubs.
COLCHICA, *kol*-chik-a, from Colchis.
COULOMBIERI, koo-lom-be-*air*-i, after Coulombier, a French nurseryman.
PINNATA, pin-*na*-ta, the leaves pinnate.
TRIFOLIA, trif-*o*-le-a, the leaves having three leaflets.

**Statice**, *stat*-is-e; from Gr. *statikos*, arresting, or to cause to stand, in reference to the supposed power of the plant in checking bleeding. Annuals and herbaceous perennials.
BONDUELLII, bon-du-*el*-lei, after Bon-duell.
BOURGÆI, boor-*je*-i, after Eugene Bour-geau, traveller and plant collector.
DUMOSA, dew-*mo*-sa, bushy.
EXIMIA, eks-*im*-e-a, excellent.
GMELINII, mel-*e*-nei, after Samuel G Gmelin, a Russian botanist.

**Statice** (*continued*)

LATIFOLIA, lat-if-*o*-le-a, broad-leaved.
LIMONIUM, lim-*o*-ne-um, Limonium (*q.v.*) The Sea Lavender.
MINUTA, min-*ew*-ta, small, minute.
OVALIFOLIA, o-val-if-*o*-le-a, oval-leafed.
PROFUSA, pro-*fu*-sa, profuse flowering.
SINUATA, sin-u-*a*-ta, leaves scalloped.
SUWOROWII, su-wor-*o*-ei, after Suworow, a Russian.
TATARICA, tah-*tar*-ik-a, of Tartary.

**Stauntonia**, staun-*to*-ne-a; after Sir George Staunton, a traveller, and once attached to Chinese embassy. Shrub.

HEXAPHYLLA, heks-af-*il*-la, usually six leaflets to a leaf.

**Stellaria**, stel-*lar*-e-a; from L. *stella*, a star, in reference to the star-shaped flowers. Rock plants.

GRAMINEA, gram-*in*-e-a, grass-like, the foliage.
HOLOSTEA, hol-*lo*-ste-a, old generic name meaning "entire" and "a bone," used in ancient medicine for healing fractures. The Stitchwort.

**Stenanthium**, sten-*an*-the-um; from Gr. *stenos*, narrow, and *anthos*, a flower, alluding to the finely cut segments of the corolla. Hardy perennial.

ROBUSTUM, ro-*bus*-tum, robust.

**Stenotaphrum**, sten-ot-*af*-rum; from Gr. *stenos*, narrow, and *taphros*, a trench— the floral spikelets being situated in cavities in the stem or rachis. Variegated grass.

SECUNDATUM VARIEGATUM, sek-un-*da*-tum var-e-eg-*a*-tum, one-sided — the flower spike, and variegated — the foliage.

**Stephanandra**, stef-an-*an*-dra; from Gr. *stephane*, a crown, and *andros*, a stamen, in allusion to the form of the latter. Shrubs.

FLEXUOSA, fleks-u-*o*-sa, flexuous or zig-zag—the branches.
TANAKAE, tan-*a*-ke, a Japanese name.

**Stephanotis**, stef-an-*o*-tis; from Gr. *stephanos*, a crown, and *otos*, an ear, referring to the auricles of the staminal crown. Tropical climbing flowering shrub.

FLORIBUNDA, flor-ib-*un*-da, abundant flowered.

**Sternbergia**, stern-*ber*-ge-a; named after Count Sternberg, a German botanist. Bulbous plants.

**Sternbergia** (*continued*)

COLCHICIFLORA, kol-tshe- (or kol-ke-) se-*flor*-a, Colchicum-flowered.
FISCHERIANA, fish-er-e-*a*-na, of Fischer, a Russian professor of botany.
GRÆCA, *gre*-ka, Grecian.
LUTEA, *lu*-te-a, yellow.
MACRANTHA, mak-*ran*-tha, large-flowered.

**Stevia**, *steve*-e-a; after P. J. Esteve, botanist of Valencia. Herbaceous perennial.

SERRATA, ser-*ra*-ta, saw-edged—the leaves.

**Stewartia**, stew-*ar*-te-a; named in honour of John Stuart, Earl of Bute, chief adviser to Augusta, Princess Dowager of Wales, when she founded the Royal Botanic Gardens, Kew, 1759-1760. Flowering shrubs.

PENTAGYNA, pen-*ta*-jin-a, five-styled.
PSEUDO-CAMELLIA, *sue*-do-kam-*el*-le-a, false camellia, its flowers resembling those of that shrub.

**Stigmaphyllon**, stig-maf-*il*-lon; from Gr. *stigma*, the receptive top of the pistil, and *phyllon*, a leaf—the stigma is leaflike or foliaceous. Tropical flowering climbing shrub.

CILIATUM, sil-e-*a*-tum, fringed with fine hairs.

**Stipa**, *sty*-pa; from L. *stuppa*, tow, in allusion to the silkiness of the flower spike. Ornamental grasses.

ELEGANTISSIMA, el-e-gan-*tis*-sim-a, most elegant.
PENNATA, pen-*na*-ta, feathered. The Feather Grass.

**Stokesia**, *stokes*-e-a; named after Dr. Stokes, an English botanist. Herbaceous perennial.

CYANEA, sy-*a*-ne-a, azure. Stokes' Aster.

**Stranvæsia**, stran-*ve*-ze-a; named by Lindley, the botanist, in honour of Mr. Fox Strangeways. Shrubs.

GLAUCESCENS, glaw-*ses*-senz, somewhat grey-leaved.
UNDULATA, un-dul-*a*-ta, waved—the leaf margins.

**Stratiotes**, strat-e-*o*-tez; from Gr. *stratiotis*, a soldier, in allusion to the sword-like leaves. Aquatic.

ALOIDES, al-*oy*-dez, aloe-like. The Water Soldier.

**Strelitzia**, stre-*lits*-e-a; after Charlotte, Queen of George III, of the House of

106

**Strelitzia** (*continued*)
Mecklenburg-Strelitz. Greenhouse perennials.

> REGINÆ, re-*ji*-ne, of the queen—Queen Charlotte.

**Streptocarpus,** strep-to-*kar*-pus; from Gr. *streptos*, twisted, and *karpos*, a fruit, the latter being spiralled. Greenhouse perennials. Many hybrids.

> DUNNII, *dun*-nei, after Dunn.
> POLYANTHA, pol-e-*an*-tha, many-flowered.
> WENDLANDII, wend-*land*-ei, after Wendland, botanist, of Hanover.

**Streptosolen,** strep-to-*so*-len, from Gr. *streptos*, twisted, and *solon*, tube, with reference to the form of the corolla tube. Greenhouse shrub.

> JAMESONII, jame-*so*-nei, of Dr. Jameson.

**Strobilanthes,** strob-il-*an*-thez; from Gr. *strobilos*, a pine cone, and *anthos*, a flower, the flower-head—especially in the bud stage—being cone-like. Greenhouse flowering plants.

> ANISOPHYLLUS, an-is-of-*il*-lus, unequal-leaved.
> DYERIANUS, dy-er-e-*a*-nus, after Dyer.
> ISOPHYLLUS, is-of-*il*-lus, equal-leaved.

**Struthiopteris,** stru-the-*op*-ter-is; from Gr. *strouthion*, an ostrich, and *pteris*, a fern, the fronds being supposed to resemble an ostrich's feather. Hardy ferns.

> GERMANICA, jer-*man*-ik-a, of Germany.

**Stuartia,** stew-*ar*-te-a; a variant spelling of Stewartia, which see.

**Stylophorum,** sty-*lo*-for-um; from Gr. *stulos*, a style, and *phoreo*, to bear, probably in allusion to the style being retained on the seed capsules. Herbaceous perennial.

> DIPHYLLUM, dif-*il*-lum, two-leaved.

**Styrax,** *sty*-raks; ancient Greek name, derived from the Arabic for a shrub, probably S. officinale, yielding the resin known as storax. Shrubs.

> JAPONICUM, jap-*on*-ik-um, of Japan.
> OBASSIA, o-*bas*-se-a, a Japanese name.
> OFFICINALE, of-fis-in-*a*-le, of the shops. The Storax.
> VEITCHIORUM, veech-e-*or*-um, introduced by Wilson for Messrs. Veitch, late of Chelsea.
> WILSONII, wil-*so*-nei, after E. H. Wilson, who introduced it.

**Sutherlandia,** suth-er-*lan*-de-a; named after James Sutherland, botanical author. Half-hardy shrubs.

> FRUTESCENS, frut-*es*-senz, shrubby.

**Swainsonia,** swain-*so*-ne-a; after Isaac Swainson, F.R.S. Greenhouse shrubs.

> GALEGIFOLIA, gal-e-gif-*ol*-e-a, Galega-like foliage.

**Swertia,** *swer*-te-a; after E. Swert, a Dutch florist. Dwarf rock and bog plants

> PERENNIS, per-*en*-nis, perennial.

**Symphoricarpus,** sim-for-e-*kar*-pus; from Gr. *symphoreo*, to accumulate, and *karpos*, a fruit, alluding to the clustered fruits. Berry-bearing shrubs.

> OCCIDENTALIS, oks-se-*den*-ta-lis, western.
> ORBICULATUS, or-bik-ul-*a*-tus, orbicular—the leaves.
> RACEMOSUS, ras-em-*o*-sus, flowers and fruits in racemes. The Snowberry.

**Symphyandra,** sim-fi-*an*-dra; from Gr. *symphyo*, to unite, and *andros*, a stamen, in reference to the united formation of the latter. Herbaceous perennials.

> HOFFMANNII, hof-*man*-ei, after Hoffmann, professor of botany.
> PENDULA, *pen*-du-la, flowers pendant.

**Symphytum,** *sim*-fit-um; from Gr. *symphyo*, to make whole, or heal, in allusion to the medicinal properties. Herbaceous perennials.

> ASPERRIMUM, as-*per*-re-mum, roughest.
> CAUCASICUM, kaw-*kas*-ik-um, Caucasian.
> OFFICINALE, of-fis-in-*a*-le, of the shops (herbal). The Comfrey.

**Symplocarpus,** sim-plo-*kar*-pus; from Gr. *symploke*, connected, and *karpos*, fruit, the seed being united in a mass. Tuberous-rooted perennial.

> FŒTIDUS, *fet*-id-us, fœtid, evil-smelling. The Skunk Cabbage.

**Synthyris,** sin-*thir*-is; from Gr. *syn*, together, and *thyris*, a small aperture, referring to the formation of the seed vessel. Rock plants.

> PINNATIFIDA, pin-na-*tif*-id-a, leaves pinnately cut.
> RENIFORMIS, ren-if-*or*-mis, kidney-shaped—the leaves.

**Syringa,** sy-*ring*-a; from Gr. *syrinx*, a tube, said to be in allusion to the hollow stems sometimes used for pipe-stems. Flowering shrubs.

> CHINENSIS, tshi-*nen*-sis, of China. The Rouen Lilac.

107

**Syringa** (*continued*)

JOSIKÆA, jos-ik-*e*-a, after Baroness von Josika, who took part in its first introduction.

OBLATA, o-*bla*-ta, widened—the broad leaves.

PEKINENSIS, pe-kin-*en*-sis, first sent from Pekin.

PERSICA, *per*-sik-a, of Persia. The Persian Lilac.

VILLOSA, vil-*lo*-sa, shaggy with hairs, of no application to the plant so named.

VULGARIS, vul-*gar*-is, common. The Common Lilac.

**Tabernæmontana**, tab-er-na-mon-*ta*-na; after James Theodore Tabernæmontanus, German physician and botanist. A large genus of tropical flowering trees and shrubs.

CORONARIA FL.-PL., kor-on-*air*-e-a (crowned) *flor*-e-*ple*-no (double flowered).

GRATISSIMA, gra-*tis*-sim-a, gratefully fragrant or very sweet scented.

**Tacsonia**, tak-*so*-ne-a; from *tacso*, a Peruvian name for one of the species. Greenhouse climbers.

EXONIENSIS, ex-o-ne-*en*-sis, of Exeter.

MANICATA, man-ik-*a*-ta, collared or sleeved.

VAN VOLXŒMII, van-volks-*e*-mei, after Van Volxem, a botanist.

**Tagetes**, ta-*ge*-tez, after L. *Tages*, an Etruscan divinity. Half-hardy annuals.

ERECTA, e-*rek*-ta, erect. The African Marigold.

PATULA, *pat*-ul-a, spreading. The French Marigold.

SIGNATA, sig-*na*-ta, notable, also means marked, as with writing. The Mexican Marigold.

**Tamarix**, *tam*-ar-iks; from Tamaris (Tambro), a river in Spain, where some species abound. Flowering and hedge shrubs.

ANGLICA, *ang*-lik-a, of England. The Tamarisk.

GALLICA, *gal*-lik-a, French.

HISPIDA, *his*-pid-a, bristly.

PENTANDRA, pent-*an*-dra, five-stamened.

TETRANDRA, tet-*ran*-dra, four-stamened.

**Tanacetum**, tan-a-*se*-tum; derivation uncertain but said to be from Gr. *athanatos*, immortal, in allusion to the long-lasting flowers. Herbaceous perennials and rock plants.

ARGENTEUM, ar-*jen*-te-um, silvery.

HERDERI, her-*deer*-i, after Herder.

**Tanacetum** (*continued*)

VULGARE, vul-*gar*-e, common. The Tansy.

V. CRISPUM, *kris*-pum, finely divided leaves.

**Tanakea**, tan-a-*ke*-a; in honour of M. Tanaka. Rock plant.

RADICANS, *rad*-e-kanz, rooting, *i.e.*, the runners.

**Taraxacum**, ta-*rax*-a-kum; from a native Arabian name. The well-known perennial weed sometimes used as a salad vegetable.

OFFICINALE, of-fis-in-*a*-le, of the shop. The Dandelion.

**Tauscheria**, towsh-*e*-re-a; after Ignatius Tauscher, a botanist, of Austria. Annual.

LASIOCARPA, las-e-o-*kar*-pa, hair-fruited.

**Taxodium**, taks-*o*-de-um, from L. *Taxus*, the yew, and Gr. *oides*, like. Coniferous trees with yew-like leaves.

DISTICHUM, *dis*-tik-um, the leaves in two rows. The Deciduous Cypress.

**Taxus**, *taks*-us; Latin name for a Yew tree; perhaps from Gr. *taxon*, a bow, the wood being once used for making bows. Evergreen coniferous trees.

BACCATA, bak-*ka*-ta, berried. The Common Yew.

B. FASTIGIATA, fas-tij-e-*a*-ta, tapering. The Irish Yew.

BREVIFOLIA, brev-if-*ol*-e-a, short-leaved.

CUSPIDATA, kus-pid-*a*-ta, the leaves tipped with a short sharp point. The Japanese Yew.

**Tchihatchewia**, popularly ke-hatch-*ew*-e-a; named after the Russian botanist Tchihatcheff. Rock plants.

ISATIDEA, i-sa-*tid*-e-a, Isatis-like.

**Tecoma**, te-*ko*-ma; said to be a contraction of the Mexican name Tecomaxochili. Greenhouse and hardy climbers.

AUSTRALIS, aws-*tra*-lis, southern.

CAPENSIS, ka-*pen*-sis, from the Cape.

GRANDIFLORA, gran-dif-*lo*-ra, large-flowered.

RADICANS, *rad*-e-kanz, trailing and rooting.

**Tecophilæa**, te-*kof*-e-le-a; after Tecophila, a daughter of Billotti Bertero, a botanist. Greenhouse flowering bulbs.

CYANOCROCUS, cy-*an*-o-kro-kus, referring to the colour (blue) and the shape (like a crocus), lit. the blue crocus.

**Telanthera,** tel-an-*the*-ra; from Gr. *teleios*, complete or perfect, and *anthera*, an anther, the latter all being of equal length. Tender plants, for carpet bedding. This genus is now included in Alternanthera.

AMABILIS, am-*a*-bil-is, lovely.

AMŒNA, am-*e*-na, lovely, pleasing.

BETTZICKIANA, bets-ik-i-*a*-na, named after Bettzick.

FICOIDEA, fi-*koy*-de-a, Ficoideæ (Fig Marigold)-like

VERSICOLOR, ver-*sik*-o-lor, various or changeable colours.

**Tellima,** tel-*li*-ma; said to be an anagram of Mitella, from which the genus was separated. Woodland perennials.

GRANDIFLORA, gran-dif-*lo*-ra, large-flowered.

**Testudinaria,** tes-tu-din-*ar*-e-a; from L. *testudo*, a tortoise, referring to the appearance of the woody root-stock. Greenhouse climber.

ELEPHANTIPES, el-ef-*an*-tip-ez, elephant's foot—another allusion to the rootstock. Hottentot Bread.

**Tetragonia,** tet-ra-*go*-ne-a; from Gr. *tetra*, four, and *gonia*, an angle, the fruits being four angled. Hardy annual grown as spinach substitute in the vegetable garden. The New Zealand Spinach.

EXPANSA, ex-*pan*-sa, spreading.

**Tetratheca,** tet-ra-*the*-ka; from Gr. *tetra*, four, *theke*, a box or cell, alluding to the anthers being four-celled. Greenhouse flowering shrubs.

ERICÆFOLIA, er-ik-e-*fol*-e-a, heath (Erica)-leaved.

HIRSUTA, hir-*su*-ta, hairy.

VERTICILLATA, ver-tis-il-*la*-ta, leaves in whorls.

**Teucrium,** *tew*-kre-um; named after Teucher, a Trojan prince who first used one of the species in medicine. Annuals, herbaceous perennials, rock plants, and shrubs.

AUREUM, *aw*-re-um, golden—the flowers.

CHAMÆDRYS, kam-*e*-dris, old name for Germander, signifying on the ground.

FRUTICANS, *frut*-ik-anz, shrubby.

LUCIDUM, *lu*-sid-um, shining—the foliage.

MARUM, *mar*-um, old name for Cat Thyme, of uncertain origin. The Cat Thyme.

MASSILIENSE, mas-sil-i-*en*-se, of Marseilles.

**Teucrium** (*continued*)

POLIUM, *po*-le-um, early name for Poly Germander, probably signifying grey —the foliage. The Poly Germander.

PYRENAICUM, pir-en-*a*-ik-um, Pyrenean.

**Thalia,** *tha*-le-a; after J. Thalius, a German scientist. Half-hardy perennials.

DEALBATA, de-al-*ba*-ta, whitened, the foliage.

**Thalictrum,** thal-*ik*-trum; old Gr. name, possibly from *thallo*, to flourish, or to abound in. referring to the numerous flowers. Herbaceous perennials.

ADIANTIFOLIUM, ad-e-an-tif-*o*-le-um, Adiantum-leaved.

ANEMONOIDES, an-em-on-*oy*-dez, anemone-like.

ALPINUM, al-*pine*-um, alpine.

AQUILEGIFOLIUM, ak-wil-e-jif-*o*-le-um, Aquilegia-leaved.

DELAVAYI, del-a-*va*-i, after Delavay, plant collector.

DIPTEROCARPUM, dip-ter-o-*kar*-pum, two-winged fruits.

FLAVUM, *fla*-vum, yellow.

GLAUCUM, *glaw*-kum, grey-blue, the foliage.

MINUS, *mi*-nus, less.

**Thelocactus,** thel-o-*kak*-tus; from Gr. *thele*, a nipple and *cactus*; refers to the tubercled ribs of the plants. Greenhouse cacti.

BICOLOR, *bik*-o-lor, two-coloured.

HEXÆDROPHORUS, hex-a-*drof*-or-us, having six sides—the tubercles.

SUBTERRANEUS, sub-ter-*ran*-e-us, refers to the tuberous root.

**Theobroma,** the-o-*bro*-ma; from Gr. *theos*, god, and *broma*, food—celestial food or food for gods, referring to the fruit—the cocoa or chocolate nut. Warm-house shrub.

CACAO, ka-*ka*-o, from the Mexican name Cacauare.

**Thermopsis,** ther-*mop*-sis; from Gr. *thermos*, a lupin, and *opsis*, a resemblance. Herbaceous perennials.

CAROLINIANA, kar-ro-li-ne-*a*-na, from Carolina.

FABACEA, fa-*ba*-se-a Faba (broad-bean)-like.

MONTANA, mon-*ta*-na, of mountains.

**Thladiantha,** thlad-e-*an*-tha; from Gr. *thladias*, compressed, and *anthos*, a flower, the subject being first described from pressed material. Tender annual climber.

DUBIA, *du*-be-a, intermediate.

**Thlaspi,** *thlas*-pe; from *thlaspis*, an old Greek name for a kind of cress. Rock plants.

    BELLIDIFOLIUM, bel-lid-if-*o*-le-um, daisy-leaved.

    ROTUNDIFOLIUM, ro-tun-dif-*o*-le-um, round-leaved.

**Thrinax,** *thrin*-aks; from Gr. *thrinax*, a fan—the shape of the leaves. Warm-house palms.

    ARGENTEA, ar-*jen*-te-a, silvery—the leaves are silky underneath.

    EXCELSA, ex-*sel*-sa, tall.

    PARVIFLORA, par-vif-*lor*-a, small flowered.

    RADIATA, rad-e-*a*-ta, rayed—the leaf divisions.

**Thuja,** classical name, possibly from Gr. *thuia*, ancient name for some resin-bearing tree; or from Gr. *thuon*, a sacrifice, at which the resin would be burned as incense. Arbor-vitae. Coniferous trees.

    DOLABRATA, dol-a-*bra*-ta, axe-shaped, presumably the leafage.

    GIGANTEA, ji-*gan*-te-a, giant.

    JAPONICA, jap-*on*-ik-a, Japanese.

    LOBBII, *lob*-bei, after Lobb.

    OCCIDENTALIS, oks-se-den-*ta*-lis, western—North America.

    ORIENTALIS, or-e-en-*ta*-lis, eastern—Eastern Asia (China).

    PLICATA, pli-*ka*-ta, folded—the leaves.

**Thujopsis,** thew-*yop*-sis; from Gr. *thuja*, thuja, and *opsis*, likeness—similar to Thuja. Also pronounced thew-e-*op*-sis. Coniferous tree.

    DOLABRATA, dol-a-*bra*-ta, axe-shaped.

**Thunbergia,** thun- (or tun-) *ber*-ge-a; named after Karl P. Thunberg, Swedish botanist, and student of Linnæus, and traveller in Japan and S. Africa. Greenhouse climbers.

    ALATA, al-*a*-ta, winged—the leaf stalks.

    ERECTA, e-*rek*-ta, erect or bushy.

    GRANDIFLORA, gran-dif-*lor*-a, large-flowered.

    HARRISII, *har*-ris-ei, after Harris.

    LAURIFOLIA, law-rif-*o*-le-a, Laurus-leaved.

    MYSORENSIS, my-sor-*en*-sis, from Mysore, India.

**Thunia,** tu-ne-a; after Count Thun-Telschen who had an important collection of orchids. Greenhouse terrestrial orchids.

    MARSHALLIANA, mar-shal-le-*a*-na, after Marshall.

**Thymus,** *ty*-mus; old Gr. name used by Theophrastus either for this plant or for savoury. Shrubby and trailing rock plants and culinary herbs.

    AZORICUS, az-*or*-ik-us, from the Azores.

    CARNOSUS, kar-*no*-sus, flesh-coloured—the leaves.

    CITRIODORUS, sit-re-od-*or*-us, citron-scented. The Lemon Thyme.

    ERECTUS, e-*rek*-tus, erect.

    HERBA-BARONA, *her*-ba-bar-*o*-na, herb baron.

    MICANS, *mik*-ans, shining, glistening.

    NITIDUS, *nit*-id-us, lustrous.

    SERPYLLUM, ser-*pil*-lum, old Gr. name (*kerpyllos*) for the Wild Thyme.

    S. LANUGINOSUS, lan-u-jin-*o*-sus, woolly.

    VULGARIS, vul-*gar*-is, common. The Garden Thyme.

**Thyrsacanthus,** thir-sak-*an*-thus, from Gr. *thyrsos*, a thyrse, and *acanthus*—thyrse-flowered acanthus. Greenhouse flowering shrubs.

    RUTILANS, *roo*-til-anz, reddish—the flowers.

**Tiarella,** te-ar-*el*-la, from Gr. *tiara*, a turban (lit. a little turban), alluding to the shape of the seed-pod. Woodland plants.

    CORDIFOLIA, kor-dif-*ol*-e-a, heart-shaped—the leaves.

    UNIFOLIATA, u-nif-ol-e-*a*-ta, one-leaved, *i.e.*, the leaves rising direct and singly from root-stock.

**Tibouchina,** tib-ou-*shy*-na; the native Guianan name. Greenhouse flowering shrubs.

    ELEGANS, *el*-e-ganz, elegant.

    SEMIDECANDRA, sem-e-dek-*an*-dra, half ten-anthered.

**Tigridia,** tig-*rid*-e-a; from Gr. *tigris*, a tiger, and *eidos*, like, referring to the brightly spotted flowers. Half-hardy bulbs.

    PAVONIA, pa-*vo*-ne-a, a peacock—the gaily-coloured blossoms.

    P. ALBA, *al*-ba, white.

    P. CONCHIFLORA, kong-kif-*lo*-ra, shell-like flowers.

**Tilia,** *til*-e-a; old Latin name for Lime tree. Trees.

    CORDATA, kor-*da*-ta, heart-shaped—the leaves.

    EUCHLORA, u-*klor*-a, dark green.

    MONGOLICA, mon-*gol*-ik-a, Mongolian.

    PETIOLARIS, pet-e-o-*lar*-is, with long leaf-stalks.

**Tilia** (*continued*)

PLATYPHYLLUS, plat-e-*fil*-lus, broad-leaved.

TOMENTOSA, to-men-*to*-sa, felted, the leaf under parts.

VULGARIS, vul-*gar*-is, common. The Lime Tree or Linden.

**Tillandsia**, til-*land*-se-a; after Elias Tillands, a Swedish cataloguer of plants in Abo, Finland. Tropical epiphytes, flowering and ornamental foliage.

ANCEPS, *an*-seps, two-edged.

DUVALIANA, du-val-e-*a*-na, after Duval.

HIEROGLYPHICA, hi-er-o-*glif*-ik-a, hieroglyphic-like markings on the leaves.

LINDENII, *lin*-den-e, after Linden.

REGINA, re-*ji*-na, queen.

TESSELLATA, tes-sel-*la*-ta, tessellated.

USNEOIDES, us-ne-*oy*-dez, like Usnea, a genus of lichens, some pendulous. The Spanish Moss.

**Tinantia**, te-*nan*-te-a; after M. Tinant, a Belgian botanist. Greenhouse or window herbaceous perennial.

FUGAX ERECTA, *fu*-gax e-*rek*-ta, fleeting (flowers) and erect (habit of plant).

**Titanopsis**, ti-tan-*op*-sis; from Gr. *titan*, the sun, and *opsis*, like, alluding to the round yellow flowers. Greenhouse succulent perennial.

CALCAREA, kal-kar-*e*-a, spurred.

**Todea**, *to*-de-a; after Henry Julius Tode, mycologist of Mecklenburg. Greenhouse ferns, coriaceous and filmy fronded.

AFRICANA, af-re-*ka*-na, of Africa.

BARBARA, *bar*-bar-a, foreign.

HYMENOPHYLLOIDES, hi-men-of-il-*loy*-dez, Hymenophyllum-like.

PELLUCIDA, pel-*lu*-sid-a, pellucid or transparent.

SUPERBA, su-*per*-ba, superb.

**Tofieldia**, to-*feel*-de-a; after Mr. Tofield, a friend of Hudson, the author of the genus. Herbaceous perennials.

PALUSTRIS, pal-*us*-tris, inhabiting marshes.

PUBENS, *pew*-bens, downy—the foliage.

**Tolmiea**, *tol*-me-a; after Mr. Tolmie, surgeon of Hudson Bay Company. Herbaceous perennial.

MENZIESII, men-*zeez*-ei, after Menzies.

**Tolpis**, *tol*-pis; derivation unexplained. Hardy annual.

BARBATA, bar-*ba*-ta, bearded.

**Torenia**, tor-*e*-ne-a; after Rev. Olaf Toren, a Swede who discovered Torenia asiatica in China, 1845. Greenhouse flowering annuals.

ASIATICA, aysh-e-*at*-ik-a, Asiatic.

BAILLONII, bail-*lo*-nei, after Baillon.

FLAVA, *fla*-va, yellow.

FOURNIERI, four-ne-*air*-i, after Fournier.

**Torreya**, tor-*re*-a; after Dr. Torrey, an American botanist. Evergreen coniferous trees.

CALIFORNICA, kal-if-*or*-nik-a, of California.

NUCIFERA, new-*sif*-er-a, nut-bearing.

**Tournefortia**, tourn-*for*-te-a; after Joseph Pitton de Tournefort (1656-1708), a notable systematic botanist whose labours laid the foundation for the "Natural System" of plant classification. Greenhouse flowering shrubs.

CORDIFOLIA, kor-dif-*ol*-e-a, heart-shaped or cordate leaves.

HELIOTROPIOIDES, he-le-o-trop-e-*oy*-dez, Heliotropium-like.

LÆVIGATA, lev-e-*ga*-ta, smooth.

**Townsendia**, town-*send*-e-a; after Townsend, a botanist. Rock plant.

WILCOXIANA, wil-koks-e-*a*-na, after Wilcox.

**Trachelium**, trak-*e*-le-um; from Gr. *trachelos*, the throat, in reference to the uses of the plant in ancient medicine. Herbaceous perennials.

CÆRULEUM, ser- (or ker-) *u*-le-um, sky-blue. The Throat-wort.

**Trachelospermum**, trak-el-o-*sper*-mum; from Gr. *trachelos*, throat, and *sperma*, seeds; application obscure. Greenhouse twining plants.

JASMINOIDES, jas-min-*oy*-dez, jasmine-like.

**Trachycarpus**, trak-e-*kar*-pus; from Gr. *trachus*, rough, and *karpus*, a fruit, the seeds of some species being hairy. Hardy palm.

EXCELSA, eks-*sel*-sa, tall—when full grown.

**Trachymene**, trak-e-*me*-ne; from Gr. *trachus*, rough, and *hymen*, a membrane—the channels in the fruit. Half-hardy annuals.

CÆRULEA, ser- (or ker-) *u*-le-a, sky-blue. The Blue Lace Flower.

**Tradescantia**, trad-es-*kan*-te-a; after John Tradescant, gardener to Charles I. Greenhouse and hardy herbaceous perennials.

REGINÆ, re-*ji*-ne, queen—Queen Victoria.
ROSEA, *ro*-ze-a, rosy.
VIRGINIANA, ver-jin-e-*a*-na, of Virginia.
VITTATA, vit-*ta*-ta, striped.
ZEBRINA, ze-*bry*-na, zebra-striped.

**Tragopogon**, trag-o-*po*-gon; from Gr. *tragos*, a goat, and *pogon*, a beard, referring to the long silky beards on the seeds. Culinary vegetable.

PORRIFOLIUS, por-rif-*ol*-e-us, leek (Allium porrum)-leaved. The Salsafy.

**Trapa**, *tra*-pa; from L. *calcitrapa*, an ancient four-pronged instrument once used in warfare for impeding navigation. Word applied here in allusion to the four-horned seeds. Aquatic perennials.

NATANS, *na*-tans, floating. The Water Caltrop or Water Chestnut.

**Tremandra**, tree-*man*-dra; from Gr. for hole and anther, the anthers burst open through a hole. Greenhouse flowering shrub.

STELLIGERA, stel-*lig*-er-a, star-bearing—refers to stellate down on the leaves.

**Trichocereus**, *trik*-o-*se*-re-us; from Gr. *trichos*, a hair, and *cereus*; the flowers have hairy tubes. Greenhouse cacti.

MACROGONUS, mak-ro-*go*-nus, large angled.
SPACHIANUS, spash-e-*a*-nus, of Spachia, Mexico.

**Trichodiadema**, *trik*-o-di-a-*dem*-a; from Gr. *trichos*, a hair, and *diadema*, to bind round. Greenhouse succulent plants.

BARBATUM, bar-*ba*-tum, bearded.
DENSUM, *den*-sum, dense (bearded).
STELLATUM, stel-*la*-tum, starry (bearded).
STELLIGERUM,s tel-*lig*-er-um, star-bearing.

**Tricholæna**, trik-ol-*e*-na; from Gr. *thrix*, a hair, and *chlaina*, a cloak, referring to the shaggy spikelets. Half-hardy annual ornamental grass.

ROSEA, *ro*-ze-a, rosy—the flower heads.

**Trichomanes**, trik-*om*-an-eez; from Gr. *thrix*, a hair, and *manos*, soft, the shining stems and soft, pellucid fronds. Greenhouse filmy ferns.

BANCROFTII, ban-*kroft*-ei, after Bancroft.
RADICANS, *rad*-e-kanz, rooting from the creeping stems. The Killarney Fern.
RENIFORME, ren-if-*or*-me, kidney-shaped—the fronds.
VENOSUM, ve-*no*-sum, veined.

**Trichopilia**, trik-op-*e*-le-a; from Gr. *thrix*, a hair, and *pilion*, a cap, in allusion to three tufts of hair surmounting the column and hiding the anther bed. Greenhouse orchids.

CRISPA, *kris*-pa, curled.
SUAVIS, *swa*-vis, sweet—the fragrance.
TORTILIS, *tor*-til-is, twisted—the petals.

**Trichosanthes**, trik-os-*an*-theez; from Gr. *thrix*, a hair, and *anthos*, a flower—the edges of the corolla limbs are ciliated or fringed. Greenhouse ornamental gourd.

ANGUINA, an-*gwee*-na, snake-like—the fruits.

**Tricuspidaria**, trik-us-pid-*ar*-e-a; from L. *tricuspis*, having three points, *i.e.*, the three teeth of the petals. Half-hardy flowering shrubs.

DEPENDENS, de-*pen*-dens, hanging down—the flowers.
LANCEOLATA, lan-se-o-*la*-ta, lance-shaped—the leaves.

**Tricyrtis**, trik-*er*-tis; from Gr. *treis*, three, and *kyrtos*, convex, referring to the three outer sepals having swollen bases. Tender perennial.

HIRTA, *her*-ta, hairy—the plant is softly hairy.

**Trientalis**, tre-en-*ta*-lis; from L. *trien*, one-third of a foot, the height of the plant. Woodland herb.

AMERICANA, a-mer-ik-*a*-na, of North America.
EUROPÆA, u-*ro*-pe-a, of Europe.

**Trifolium**, trif-*ol*-e-um; from L. *tres*, three, and *folium*, a leaf—the trefoil leafage. Trailing and other perennials and annuals.

ALPINUM, al-*pine*-um, alpine.
BADIUM, *ba*-de-um, chestnut-brown.
REPENS PURPUREUM, *re*-penz pur-*pur*-e-um, creeping and purple—the leaves.
RUBENS, *roo*-bens, red, usually means dark red.
UNIFLORUM, u-nif-*lo*-rum, one-flowered.

**Trillium**, *tril*-le-um; from L. *triplum*, triple, alluding to the three-parted flowers. Tuberous perennials. The Wood Lilies.

ERECTUM, e-*rek*-tum, erect.
GRANDIFLORUM, grand-dif-*lor*-um, large-flowered.
OVATUM, o-*va*-tum, egg-shaped, presumably the flower segments.

**Trillium** (*continued*)
RECURVATUM, re-kur-*va*-tum, recurved—the flowers.
RIVALE, re-*va*-le, growing by brook-sides.
SESSILE, *ses*-sil-e, stemless, the flowers.
STYLOSUM, sty-*lo*-sum, long-styled.

**Triteleia,** trit-el-*i*-a; from Gr. *treis*, three, and *teleios*, complete, referring to the three-parted form of flowers and fruit. Hardy bulbs.
UNIFLORA, u-nif-*lo*-ra, one-flowered—per stem.

**Tritoma,** *trit*-o-ma; from Gr. *treis*, three, and *temno*, to cut, having three sharp edges at the ends of the leaves, or Gr. *tritomos*, thrice cut. This name dropped in favour of Kniphofia, which see.
UVARIA, u-*var*-e-a, a cluster.

**Tritonia,** tri-*to*-ne-a; from Gr. *triton*, a weathercock, in allusion to the variable positions of the anthers. Half-hardy bulbous plants.
CROCATA, kro-*ka*-ta, saffron-yellow.
CROCOSMÆFLORA, kro-kos-me-*flor*-a, crocosmea-flowered.
POTTSII, *potts*-ei, after Potts.

**Trochodendron,** trok-o-*den*-dron; from Gr. *trochos*, a wheel, and *dendron*, a tree, the flowers being rayed like the spokes of a wheel. Evergreen shrub.
ARALIOIDES, ar-a-le-*oy*-dez, aralia-like.

**Trollius,** *trol*-le-us; from a Swiss-German vernacular name; some authorities give German *trol*, a globe, or something round. The Globe-flowers.
ALTAICUS, al-*ta*-ik-us, from Altai Mountains.
ASIATICUS, aysh-e-*at*-ik-us, of Asia.
EUROPÆUS, u-*ro*-pe-us, of Europe.
LEDEBOURII, led-e-*bour*-ei, after Ledebour.
PATULUS, *pat*-u-lus, spreading—the sepals outspread.
POLYSEPALUS, pol-e-*sep*-a-lus, many sepals.
PUMILUS, *pew*-mil-us, dwarf.

**Tropæolum,** trop-e-*o*-lum; from L. *tropæum* (Gr. *tropaion*), a trophy, probably in allusion to the likeness of the flowers and leaves to the helmets and shields once displayed in Greece and Rome about scenes of victory. Annuals and herbaceous perennials.
ADUNCUM, ad-*unk*-um, hooked—the flowers.

**Tropæolum** (*continued*)
CANARIENSE, ka-nar-e-*en*-se, canary—the colour and shape of flowers. The Canary Creeper.
LOBBIANUM, lob-be-*a*-num, after Lobb, plant collector.
MAJUS, *ma*-jus, great. The Climbing Nasturtium.
MINUS, *mi*-nus, small. The Dwarf Nasturtium.
PENTAPHYLLUM, pen-ta-*fil*-lum, five-leaved, or divided into five.
PEREGRINUM, per-e-*gry*-num, foreign or wandering, here probably means straggling in growth.
SPECIOSUM, spes-e-*o*-sum, showy.
TUBEROSUM, tew-ber-*o*-sum, tuberous.

**Tsuga,** *su*-ga; Japanese name for T. Sieboldii. Coniferous trees.
ALBERTIANA, al-bert-e-*a*-na, after Albert.
BRUNONIANA, bru-no-ne-*a*-na, after Dr. Brown.
CANADENSIS, kan-a-*den*-sis, of Canada.
MERTENSIANA, mer-ten-se-*a*-na, after Mertens, a German botanist.
SIEBOLDII, se-*bold*-ei, after Siebold.

**Tulipa,** *tew*-lip-a; a corruption of the Persian word *thoulyban*, or *tulipant*, a turban, which the flower of the tulip is supposed to resemble. Hardy bulbs. The Tulip; most of the tulips in cultivation are hybrids.
BIFLORA, bif-*lo*-ra, two-flowered.
CLUSIANA, klew-ze-*a*-na, of Clusius, the botanist.
DASYSTEMON, das-e-*ste*-mon, with hairy stamens.
ELEGANS, *el*-e-ganz, elegant.
FLAVA, *fla*-va, yellow.
GESNERIANA, ges-ner-e-*a*-na, after Gesner, a botanist.
GREIGII, *greg*-ei; after M. G. Greig, a patron of botany.
KAUFMANNIANA, kowf-man-ne-*a*-na, of Kaufmann, a Russian botanist.
ORPHANIDEA, or-fan-*id*-e-a, Orphanidean, Greece.
PERSICA, *per*-sik-a, Persian.
PRÆCOX, *pre*-koks, early.
PRÆSTANS, *pre*-stans, excelling, standing out.
SPRENGERI, spreng-*er*-i, after Sprenger, a botanist.
SUAVEOLENS, *swa*-ve-ol-enz, sweetly scented.
SYLVESTRIS, sil-*ves*-tris, wild, appertaining to woods.
VITELLINA, vit-el-*le*-na, the colour of the yolk of an egg.

113

**Tunica,** tew-nik-a; from L. *tunica*, a coat, in reference to the character of the calyx. Hardy perennial.

SAXIFRAGA, saks-e-*fra*-ga, old generic name; now used for Rockfoils or Saxifrages; see Saxifraga.

**Tweedia,** *tweed*-e-a; after James Tweedie, plant collector in South America. Greenhouse flowering twiners.

CÆRULEA, ser- (or ker-) *u*-le-a, sky-blue.

**Tydæa,** ty-*de*-a; after Tydeus, son of Œneus, King of Caledon. Warm-house herbaceous flowering and ornamental foliage plants. Many florists' varieties.

AMABILIS, am-a-bil-is, lovely.

**Tussilago,** tus-sil-a-go; from Gr. *tussis*, a cough, its use in ancient medicine. Herbaceous perennials.

FARFARA VARIEGATA, far-*far*-a var-e-eg-*a*-ta, variegated-leaved farfara or colts-foot.

**Typha,** *ti*-fa; ancient name, possibly from Gr. *typhe*, a cat's tail, or *typhos*, a fen, the usual habitat of the reed-maces. Aquatics.

ANGUSTIFOLIA, an-gus-tif-*o*-le-a, narrow-leaved.

LATIFOLIA, lat-if-*o*-le-a, broad-leaved. The Reed Mace.

LAXMANNI, lax-man-ne, after Laxmann.

MINIMA, *min*-e-ma, smallest.

**Ulex,** *yu*-leks; ancient L. name, by some referred to the word *ulex*, used by Pliny for a similar spiny shrub. Thorny flowering shrubs.

EUROPÆUS FLORE-PLENO, u-*ro*-pe-us *flor*-e-*ple*-no, European and double-flowered. The Double-flowered Gorse.

GALLI, *gal*-li, of France (Gaul).

NANUS, *na*-nus, dwarf.

**Ulmus,** *ul*-mus; old Latin name for an Elm tree; perhaps from the Celtic name, *ulm*. Trees.

ALATA, al-*a*-ta, winged—branches having corky wings.

CAMPESTRIS, kam-*pes*-tris, of fields. The Common Elm.

FULVA, *ful*-va, tawny—the buds. The Slippery Elm.

MONTANA, mon-*ta*-na, mountain. The Wych-elm.

NITENS, *nit*-enz, shining or glossy—the leaves.

PUMILA, *pew*-mil-a, small.

STRICTA, *strik*-ta, upright, or columnar. The Cornish Elm.

VIMINALIS, vim-in-*a*-lis, twiggy.

**Umbellularia,** um-bel-lu-*lar*-e-a; from L. *umbellula*, a little shade, in allusion to the flowers being in umbels, *i.e.*, parasol-shaped clusters. Shrubs.

CALIFORNICA, kal-if-*or*-nik-a, of California.

**Umbilicus,** um-bil-*i*-kus; from L. *umbilicus*, the navel, referring to the rounded, concave leaves of some species. Succulent rock plants.

PENDULINUS, pen-dul-e-nus, drooping.

**Uniola,** *yu*-ne-o-la; ancient L. name. diminutive of *unio*, unity. Ornamental grasses.

PANICULATA, pan-ik-ul-*a*-ta, panicled.

**Urceolina,** ur-se-o-*le*-na; from L. *urceolus*, a small cup or pitcher, alluding to the small size of the membraneous floral cup or nectary. Greenhouse bulbs.

PENDULA, *pen*-du-la, pendulous—the flowers.

**Urospermum,** u-ro-*sperm*-um; from Gr. *oura*, a tail, and *sperma*, a seed, the latter having a tail-like protuberance. Border and rock plants.

DALECHAMPII, da-le-*shamp*-ei, after Dalechamp, a French botanist.

PICROIDES, pik-*roy*-dez, resembling an ox-tongue.

**Ursinia,** ur-*se*-ne-a; after John Ursinus of Regensburg (died 1666). Half-hardy annuals.

ANETHOIDES, an-eth-*oy*-dez, Anethum-like.

ANTHEMOIDES, an-them-*oy*-dez, Anth-emis-like.

PULCHRA, *pul*-kra, pretty.

**Utricularia,** u-trik-u-*lair*-e-a; from L. *utriculus*, a small bladder, in allusion to the little sacs attached to the submerged leaves. Aquatic.

VULGARIS, vul-*gar*-is, common. The Bladderwort.

**Uvularia,** u-vu-*lar*-e-a; from L. *uvula*, the palate, because of the hanging flowers. Hardy perennial.

GRANDIFLORA, gran-dif-*lo*-ra, large-flowered.

PERFOLIATA, per-fol-e-*a*-ta, perfoliate leaved—the leaves pierced by the stem.

**Vaccinium,** vak-*sin*-e-um; ancient L. name of the blueberry. Shrubs.

ARCTOSTAPHYLOS, ark-tos-*taf*-il-os, old generic name meaning fruits eaten by bears.

**Vaccinium** (*continued*)

CORYMBOSUM, kor-im-*bo*-sum, corymbed —application obscure.

ERYTHROCARPUM, er-ith-ro-*kar*-pum, red-fruited.

GLAUCO-ALBUM, *glaw*-ko-*al*-bum, blue-white—the leaf underparts.

MORTINIA, mor-*tin*-e-a, old generic name from "Mortina," by which the plant is known in Ecuador.

MYRTILLUS, mir-*til*-lus, myrtle, old name for Bilberry. The Bilberry or Whortleberry.

OVATUM, o-*va*-tum, egg-shaped—the leaves.

PENNSYLVANICUM, pen-sil-*van*-ik-um, of Pennsylvania.

ULIGINOSUM, u-lij-in-*o*-sum, growing in swamps.

VITIS-IDÆA, *vi*-tis-i-*de*-a, old generic name, meaning vine of Mount Idæa (Crete). The Cowberry.

**Valeriana,** va-leer-e-*a*-na; from L. *valeo*, to be healthy (ancient medicine); others derive word from Valerius, a Roman physician, who, it is said, first used the plant as a drug. Herbaceous perennials.

CELTICA, *kel*-tik-a, Celtic.

DIOICA, di-*oy*-ka, diœcious.

PHU, few, old name for Valerian, or Cretan Spikenard; said to mean evil-smelling.

SUPINA, su-*pi*-na, prostrate.

**Valerianella,** va-leer-e-a-*nel*-la; a diminutive of Valeriana (which see). An annual salad vegetable.

OLITORIA, ol-it-*o*-re-a, culinary. The Corn Salad or Lamb's Lettuce.

**Vallisneria,** val-lis-*neer*-e-a; after A. Vallesneri, an Italian botanist. Submerged aquatic.

SPIRALIS, spi-*ra*-lis, spiral, or coiled, the growths.

**Vallota,** val-*lo*-ta; after A. Vallot, a French botanist. Greenhouse bulbs.

PURPUREA, pur-*pur*-e-a, purple.

**Vancouveria,** van-koo-*veer*-e-a; after Capt. G. Vancouver, English explorer. Woodland herb.

HEXANDRA, heks-*an*-dra, six-stamened.

**Vanda,** *van*-da; the Sanskrit (Hindu) name of the first species introduced. Tropical orchids.

CÆRULEA, ser- (or ker-) *u*-le-a, sky-blue.

KIMBALLIANA, kim-bal-le-*a*-na, after W. S. Kimball.

SUAVIS, *swa*-vis, sweet.

**Vanda** (*continued*)

TERES, *ter*-ees, terete—the leaves and stems.

TRICOLOR, *trik*-o-lor, three-coloured.

**Vanilla,** va-*nil*-la; from Spanish *vaynilla*, a diminutive of *vaina*, signifying sheath, and bestowed because of the cylindrical seed pods suggesting the sheath of a knife. Tropical climbing orchid. The seed pods of V. planifolia are the vanilla of commerce.

PLANIFOLIA, plan-if-*ol*-e-a, flat-leaved.

**Veltheimia,** vel-*tym*-e-a; after Count Veltheim, a German botanist. Greenhouse flowering bulbs.

GLAUCA, *glaw*-ka, glaucous.

VIRIDIFOLIA, ver-id-if-*ol*-e-a, green-leaved (not glaucous).

**Venidium,** ven-*id*-e-um; name of obscure application. S. African half-hardy annuals and perennials.

CALENDULACEUM, kal-en-du-*la*-se-um, resembling Calendula.

FASTUOSUM, fas-tu-*o*-sum, stately.

**Veratrum,** ver-*a*-trum; ancient name of hellebore. False Hellebore. Herbaceous perennials.

ALBUM, *al*-bum, white—the flowers.

NIGRUM, *ni*-grum, black—the dark purple flowers.

VIRIDE, *ver*-id-e, green—the flowers.

**Verbascum,** ver-*bas*-kum; classical L. name, possibly a corruption of L. *Barbascum*, a hairy plant (*barba*, a beard), many of the mulleins having downy foliage.

CHAIXII, *schay*-zei, after Chaix, a French cleric and botanist.

CUPREUM, *ku*-pre-um, copper-coloured —the flowers.

OLYMPICUM, ol-*im*-pik-um, of Olympia.

PANNOSUM, pan-*no*-sum, roughly hairy, like woolly cloth.

PHLOMOIDES, flo-*moy*-des, like Phlomis,

PHŒNICEUM, fen-*ik*-e-um, reddish-scarlet.

THAPSIFORME, thaps-if-*or*-me, resembling Thapsus (Verbascum Thapsus).

THAPSUS, *thap*-sus, after Thapsus in ancient Africa (now Tunisia), or after Grecian Thapsos Island.

**Verbena,** ver-*be*-na; ancient L. name of the common European vervain, V. officinalis; some authorities say it is derived from L. *verbenae*, the sacred branches of olive, laurel, and myrtle.

**Verbena** (*continued*)

Half-hardy perennials frequently grown as annuals and used as bedding plants. Most of the garden verbenas are hybrids.

BONARIENSIS, bon-ar-e-*en*-sis, of Bonaria, Buenos Ayres.

CHAMÆDRIFOLIA, kam-e-drif-*o*-le-a, Germander-leaved.

ERINOIDES, er-in-*oy*-dez, Erinus-like.

HORTENSIS, hor-*ten*-sis, belonging to gardens.

TENERA, *ten*-er-a, tender, or soft to the touch.

VENOSA, ven-*o*-za, conspicuously or strongly veined.

**Verbesina**, ver-be-*se*-na; origin obscure, possibly altered from Verbena as there is some similiarity in the foliage. Half-hardy perennials.

ENCELIOIDES, en-se-le-*oy*-dez, Encelia-like.

HELIANTHOIDES, he-le-anth-*oy*-dez, Helianthus-like.

**Vernonia**, ver-*no*-ne-a; after W. Vernon, a botanical traveller. Hardy perennials.

ALTISSIMA, al-*tis*-sim-a, tallest.

ARKANSANA, ar-kan-*sa*-na, of Arkansas.

NOVÆBORACENSIS, no-ve-bor-a-*sen*-sis, from the district of New York.

**Veronica**, ver-*on*-ik-a; origin doubtful. Some authorities give it as a corruption of Betonica, the foliage of the plants being very similar, others suggest it is a Latin form of the Greek word *Beronike*, yet others refer it to the Gr. *hiera eicon*, sacred image, or the Arabic *viroo nikoo*, beautiful remembrance. Annuals, herbaceous perennials, aquatics, and shrubs.

AMPLEXICAULIS, am-pleks-e-*kaw*-lis, stem clasping leaves.

ANAGALLIS, an-a-*gal*-lis, old name for Water Pimpernel.

ANDERSONII VARIEGATA, an-der-*so*-nei var-e-eg-*a*-ta, after Anderson and with variegated leaves.

ANGUSTIFOLIA, an-gus-tif-*o*-le-a, narrow-leaved.

ARMSTRONGII, arm-*strong*-ei, after Armstrong.

BALFOURIANA, bal-four-e-*a*-na, after Professor Balfour, of Edinburgh.

BECCABUNGA, bek-ka-*bung*-a, old name for Brooklime, meaning "mouthsmart," *i.e.*, pungent.

BIDWILLII, bid-*wil*-lei, after Bidwill.

BUXIFOLIA, buks-if-*o*-le-a, Buxus (box)-leaved.

CANESCENS, kan-*es*-senz, hoary.

CARNOSULA, kar-*nos*-u-la, somewhat fleshy—the foliage.

**Veronica** (*continued*)

CATARRACTÆ, kat-ar-*rak*-te, appears to be from a local name, its flower sprays suggesting a waterfall.

CHAMÆDRYS, kam-*e*-dris, old name for Germander Speedwell, signifying "on the ground," and "an oak," the plant's habit and shape of leaves.

CHATHAMICA, chat-*ham*-ik-a, of Chatham Islands.

CINEREA, sin-er-*e*-a, ash-coloured—the foliage.

COLENSOI, kol-*en*-so-i, after Colenso.

CUPRESSOIDES, ku-pres-*oy*-dez, Cypress-like—the foliage.

DARWINIANA, dar-win-e-*a*-na, after Darwin, the naturalist.

DECUMBENS, de-*kum*-benz, lying down.

ELLIPTICA, el-*lip*-tik-a, elliptic, the leaves.

FILIFORMIS, fil-if-*or*-mis, thread-like, the growths.

GENTIANOIDES, jen-te-an-*oy*-dez, Gentian-like—the tufted foliage.

HECTORII, hek-*tor*-ei, after Hector.

HULKEANA, hul-ke-*a*-na, probably of native origin.

INCANA, in-*ka*-na, hoary.

LONGIFOLIA, long-if-*ol*-e-a, long leaved.

LYALLII, ly-*al*-ei, after Lyall.

LYCOPODIOIDES, li-ko-pod-e-*oy*-dez, resembling Lycopodium.

ORIENTALIS, or-e-en-*ta*-lis, eastern.

PARVIFLORA, par-vif-*lor*-a, small-flowered.

PECTINATA, pek-tin-*a*-ta, comb-shaped—the leaves.

PIMELIOIDES, py-mel-e-*oy*-dez, Pimelea-like.

PINGUIFOLIA, pin-gwe-*fo*-le-a, the leaves appearing fatty or greasy.

REPENS, *re*-penz, creeping.

SALICIFOLIA, sal-is-if-*o*-le-a, willow-leaved.

SALICORNIOIDES, sal-ik-or-ne-*oy*-dez, Salicornia-like.

SAXATILIS, saks-*a*-til-is, rock-haunting.

SPECIOSA, spes-e-*o*-sa, showy.

SPICATA, spe-*ka*-ta, spiked—the inflorescences.

SUBSESSILIS, sub-*ses*-sil-is, partially sitting—the leaves having little or no stalk.

TEUCRIUM, *tew*-kre-um, old generic name, which see.

TRAVERSII, tra-*ver*-sei, after Travers.

VERNICOSA, ver-nik-*o*-sa, shining as if varnished.

VIRGINICA, ver-*jin*-ik-a, of Virginia.

**Vesicaria**, ves-e-*kair*-e-a; from L. for bladder, alluding to the inflated seed pods. Branched annual or perennial herbs.

ARCTICA, *ark*-tik-a, arctic.

**Vesicaria** (*continued*)

GNAPHALIOIDEZ, naf-al-e-*oy*-des, gnaphalium-like.

GRACILIS, *gras*-il-is, slender.

UTRICULATA, u-trik-ul-*a*-ta, bladdered, the seed pods.

**Viburnum**, vi-*bur*-num; old Latin name for V. Lantana, the Wayfaring tree. Shrubs.

ACERIFOLIUM, a-ser-if-*o*-le-um, maple (Acer)-leaved.

CARLESII, *kar*-les-ei, after W. R. Carles.

DAVIDII, *da*-vid-ei, after David, missionary and plant introducer.

FRAGRANS, *fra*-granz, fragrant.

HARRYANUM, har-re-*a*-num, after Sir Harry Veitch.

HENRYI, *hen*-re-i, after Dr. Augustine Henry.

HESSEI, *hes*-se-i, after Hesse, a Hanoverian who introduced it.

HUPEHENSE, hu-pe-*en*-se, from Hupeh, China.

KANSUENSE, kan-su-*en*-se, from Kansu, China.

LANTANA, lan-*ta*-na, ancient name of Viburnum. The Wayfaring Tree.

MACROCEPHALUM, mak-ro-*sef*-a-lum, large-leaved.

ODORATISSIMUM, od-or-a-*tis*-sim-um, sweetest-scented.

OPULUS, *op*-ul-us, once the generic name for the Guelder Rose; some derive word from old L. name for Maple —the leaves of both being somewhat similar. The Guelder Rose.

O. STERILIS, *ster*-il-is, sterile. The Snowball Tree.

PRUNIFOLIUM, pru-nif-*o*-le-um, plum-leaved.

RHYTIDOPHYLLUM, ry-tid-of-*il*-lum, leaves deeply grooved.

TINUS, *ti*-nus, the old Latin name for the species, meaning obscure. The Laurustinus.

TOMENTOSUM, to-men-*to*-sum, felted—the leaves.

UTILE, *yu*-til-e, useful.

**Vicia**, *vis*-e-a; classical L. name, possibly from L. *vincio*, to bind, in reference to the clinging tendrils of many of the vetches. Annual and herbaceous perennials.

ARGENTEA, ar-*jen*-te-a, silvery, the foliage.

CRACCA, *krak*-a, old name for Tufted Vetch, possibly from Gr. *arachon*, a pea-like plant.

FABA, *fa*-ba, Faba—one-time generic name. The Broad Bean.

OROBUS, *or*-o-bus, once the generic name (*q.v.*) of the Bitter Vetch.

PYRENAICA, pir-en-*a*-ik-a, Pyrenean.

**Vicia** (*continued*)

UNIJUGA, u-ne-*ju*-ga, in single pairs, the leaves.

**Victoria**, vik-*tor*-e-a; after Queen Victoria. Tropical aquatic.

REGIA, *re*-je-a, royal.

**Vieusseuxia**, vu-*suz*-e-a; after M. Vieusseux, a Swiss botanist of Geneva. Half-hardy bulbs.

GLAUCOPIS, glau-*kop*-is, grey-eyed.

PAVONIA, pa-*vo*-ne-a, peacock. The Peacock Iris.

**Villarsia**, vil-*lar*-se-a; after Villars, a French botanist. Herbaceous perennials and aquatics.

CHILENSIS, chil-*en*-sis, Chilean.

NYMPHÆOIDES, nim-fe-*oy*-dez, Nymphæa-like.

OVATA, o-*va*-ta, egg-shaped—the leaves.

RENIFORMIS, ren-if-*or*-mis, kidney-shaped —the leaves.

**Vinca**, *vin*-ka; old L. name, possibly from L. *vincio*, to bind, alluding to the long, tough runners. Greenhouse and hardy shrubby plants.

DIFFORMIS, dif-*for*-mis, of unusual form, probably alluding to the shape of flower segments.

HERBACEA, her-*ba*-se-a, herbaceous.

MAJOR, *ma*-jor, greater.

MINOR, *mi*-nor, lesser. The Lesser Periwinkle.

ROSEA, *ro*-ze-a, rosy.

**Viola**, *vi*-o-la; the ancient Latin name for a violet (akin to Gr. *ion*, a violet).

ARENARIA, ar-en-*air*-e-a, of sandy places.

BIFLORA, bif-*lo*-ra, two-flowered, *i.e.*, in pairs.

BOSNIACA, bos-ne-*a*-ka, of Bosnia.

CALCARATA, kal-kar-*a*-ta, spurred.

CANADENSIS, kan-a-*den*-sis, Canadian.

CANINA, kan-*i*-na, dog, probably signifying common or worthless, that is, scentless. The Dog Violet.

CENISIA, sen-*is*-e-a, from Mount Cenis.

CORNUTA, kor-*new*-tq, horned—the sepals are awl-shaped and suggest horns.

CUCULLATA, kuk-ul-*a*-ta, hooded—the upper petals bent over.

DICHROA, dik-*ro*-a, two-coloured.

GRACILIS, *gras*-il-is, slender.

HEDERACEA, hed-er-*a*-se-a, ivy (Hedera)-like—the shape of the leaves.

HIRTA, *hir*-ta, hairy.

LUTEA, *lu*-te-a, yellow.

MUNBYANA, mun-be-*a*-na, after Munby.

ODORATA, od-or-*a*-ta, sweet-scented. The Sweet Violet.

PEDATA, ped-*a*-ta, footed—the bird's-claw-like leaves.

**Viola** (*continued*)

PEDUNCULATA, ped-ungk-ul-*a*-ta, ped-uncled—the long flower-stalks.

PINNATA, pin-*na*-ta, pinnated—the divided leaves.

ROTHOMAGENSIS, roth-o-ma-*gen*-sis, of Rouen.

TRICOLOR, *trik*-o-lor, three-coloured. The Heart's-ease.

**Viscaria**, vis-*kair*-e-a; from L. *viscum*, birdlime. Rock and border plants.

OCULATA, ok-ul-*a*-ta, eyed.

**Viscum**, *vis*-kum; Latin name for Mistletoe. Parasitic shrubby plant,

ALBUM, *al*-bum, white—the berries. The Mistletoe.

**Vitex**, *vi*-teks; ancient L. name of V. Agnus-castus. Hardy and greenhouse flowering shrubs.

AGNUS-CASTUS, *ag*-nus-*kas*-tus, chaste-lamb-tree—a classical name.

**Vitis**, *vi*-tis; old Latin name for Vine, perhaps from Celtic *gwyd* (pronounced *vid*), the most exalted of trees. Climbing foliage and fruiting plants.

ÆSTIVALIS, es-tiv-*a*-lis, summer.

AMURENSIS, am-oor-*en*-sis, from Amur.

ARIZONICA, ar-iz-*on*-ik-a, from Arizona.

ARMATA, ar-*ma*-ta, armed—young shoots bristly.

CANDICANS, *kan*-dik-ans, white—the shoots.

COIGNETIÆ, koyn-*et*-e-e, first introduced from Japan to France by a Madame Coignet.

FLEXUOSA, fleks-u-*o*-sa, flexuous, the growths curving in a wavy manner.

HENRYANA, hen-re-*a*-na, after Dr. A. Henry.

HETEROPHYLLA, het-er-of-*il*-la, variable in leaf form.

INCONSTANS, in-*kon*-stanz, leaves of inconstant character.

MONTICOLA, mon-*tik*-o-la, growing on hills.

QUINQUEFOLIA, kwin-kwe-*fo*-le-a, leaflets in fives. The Virginia Creeper.

SINENSIS, sin-*en*-sis, of China.

STRIATA, stri-*a*-ta, fluted or grooved—presumably the angled stems.

THUNBERGII, thun-*berg*-ei, after Thunberg, the botanist.

VINIFERA, vin-*if*-er-a, wine-producing. The Grape Vine.

VITACEA, vit-*a*-se-a, pertaining to vines.

**Vittadenia**, vit-ta-*din*-e-a; after Dr. C. Vittadini, an Austrian who wrote on n gi in early nineteenth century. Peren-

**Vittadinia** (*continued*)

nial herbs with thick rootstocks, or branching sub-shrubs.

AUSTRALIS, aus-*tra*-lis, southern.

**Vriesia**, vre-ze-a or vreez-e-a; after Dr. W. de Vriese, professor of botany at Amsterdam. Tropical flowering and foliage plants.

CARINATA, kar-in-*a*-ta, keeled.

DUVALIANA, du-val-e-*a*-na, after Duval.

PSITTACINA, sit-tak-*e*-na, parrot-like—the floral colouring.

SPLENDENS, *splen*-denz, splendid.

TESSELLATA, tes-sel-*la*-ta, tessellated or chequered.

**Wahlenbergia**, wah- (or vah-) len-*ber*-ge-a; named after Dr. Wahlenberg of Upsala, a botanical author. Rock-garden perennials.

DALMATICUS, dal-*mat*-ik-us, of Dalmatia.

HEDERACEA, hed-er-*a*-se-a, ivy (Hedera)-like—the leaves.

KITAIBELII, kit-a-*bel*-ei, after Kitaibel.

PUMILIO, pew-*mil*-e-o, rather dwarf.

SERPYLLIFOLIUS, ser-pil-if-*o*-le-us, thyme-leaved.

TASMANICA, taz-*man*-ik-a, of Tasmania.

VINCÆFLORA, vin-ke-*flor*-a, Vinca- or periwinkle-flowered.

**Waitzia**, *wait*-ze-a; after F. A. C. Waitz of Java. Half-hardy annuals.

AUREA, *aw*-re-a, golden

GRANDIFLORA, gran-dif-*lo*-ra, large flowered.

**Waldsteinia**, wald- (or vald-) *sti*-ne-a; after F. A. Waldstein-Wartenburg, German botanist. Trailing perennials.

FRAGARIOIDES, fraj-air-e-*oy*-dez, strawberry-like—the leaves.

TRIFOLIA, trif-*o*-le-a, leaves having three leaflets.

**Watsonia**, wat-*so*-ne-a; named after Sir William Watson, an English botanist. Half-hardy shrubs.

ARDERNEI, ar-*der*-nei, after Arderne.

IRIDIFOLIA, i-rid-if-*o*-le-a, Iris-leaved.

MERIANA, meer-e-*a*-na, after Sybilla Merian, a Dutch naturalist.

**Wedelia**, wed- (or ved-) *e*-le-a; after Geo. W. Wedel, a German botanist. Half-hardy perennials.

AUREA, *aw*-re-a, golden.

RADIOSA, rad-e-*o*-sa, rayed.

**Weigela**, we- (or ve-) *je*-la; after C. F. Weigel, a professor of botany (1748-1831), of Griefswald. Flowering shrubs

ROSEA, *ro*-ze-a, rosy—the floral colour.

**Weigelia,** we- (or ve-) *je*-le-a; a variant spelling of Weigela, which see.

**Weinmannia,** ween- (or veen-) *man*-ne-a; after J. W. Weinmann, a German botanist. Greenhouse shrubs.

RETICULATA, ret-ik-ul-*a*-ta, netted, or lined.
TRICHOSPERMA, trik-o-*sper*-ma, hairy-seeded.

**Wellingtonia,** wel-ling-*to*-ne-a; named in honour of the first Duke of Wellington. Coniferous trees.

GIGANTEA, ji-*gan*-te-a, gigantic. The Giant Tree of California.

**Welwitschia,** wel- (or vel-) *vits*-ke-a; after Dr. Frederic Welwitsch, a botanical traveller, who introduced this plant curiosity to Europe.

MIRABILIS, mir-*a*-bil-is, wonderful.

**Whitlavia,** whit-*la*-ve-a; named after F. Whitlaw, an Irish botanist. Annuals.

GRANDIFLORA, gran-dif-*lo*-ra, large-flowered.

**Widdringtonia,** wid-ring-*to*-ne-a; after Captain Widdrington. Greenhouse conifers.

CUPRESSOIDES, ku-pres-*soy*-dez, Cupressus (cypress)-like.
JUNIPEROIDES, ju-nip-er-*oy*-dez, Juniperus (juniper)-like.

**Wigandia,** wig- (or vig-) *an*-de-a; after Johannes Wigand, a Bishop of Pomerania. Summer bedding foliage plants.

CARACASANA, kar-*ak*-as-a-na, of Caracas.
URENS, *u*-rens, stinging.
VIGIERI, *vig*-e-air-e, after Vigier.

**Wilcoxia,** wil-*kox*-e-a; after General Timothy E. Wilcox, an American patron of botany. Greenhouse cacti.

TUBEROSUS, tu-ber-*o*-sus, tuberous-rooted.

**Wisteria,** wis-*teer*-e-a; after Caspar Wistar, professor at University of Penn. Sometimes, but not originally spelt Wistaria. Climbing flowering shrubs.

CHINENSIS, tshi-*nen*-sis, Chinese.
JAPONICA, jap-*on*-ik-a, of Japan.
MULTIJUGA, mul-te-*ju*-ga, many-paired, the leaflets.

**Witsenia,** wit-*see*-ne-a; after Nicholas Witsen, a Dutch patron of botany. Greenhouse flowering shrubs.

CORYMBOSA, kor-im-*bo*-sa, corymbose.

**Woodsia,** *wood*-se-a; after Joseph Woods, English architect and botanist. Greenhouse and hardy ferns.

HYPERBOREA, hi-per-*bor*-e-a, northern.
ILVENSIS, il-*ven*-sis, from Ilva (Elba).
OBTUSA, ob-*tu*-sa, obtuse.
POLYSTICHOIDES, pol-is-tik-*oy*-dez, polystichum-like.

**Woodwardia,** wood-*ward*-e-a; after T. J. Woodward, an English botanist. Greenhouse ferns.

AREOLATA, ar-e-o-*la*-ta, divided into open spaces between the veins.
RADICANS, *rad*-e-kanz, creeping and rooting.
VIRGINICA, vir-*jin*-ik-a, Virginian.

**Wulfenia,** woolf-*e*-ne-a; after F. Xavier Wulfen, a botanical professor, of Klagenfurt. Herbaceous perennials.

AMHERSTIANA, am-*herst*-e-a-na, after Lady Amherst.
CARINTHIACA, kar-in-the-*a*-ka, from Carinthia.

**Xanthoceras,** zanth-*os*-er-as; from Gr. *xanthos*, yellow, and *keras*, a horn, the projecting glands between the petals. Flowering tree.

SORBIFOLIA, sor-bif-*o*-le-a, Sorbus-leaved

**Xeranthemum,** zer-*an*-them-um; from Gr. *xeros*, dry, and *anthos*, a flower, alluding to the chief characteristic of these "Everlastings." Hardy annuals.

ANNUUM, *an*-nu-um, annual.

**Xerophyllum,** zer-o-*fil*-lum; from Gr. *xeros*, dry, and *phullon*, a leaf, the leaves being dry and grassy. Herbaceous perennials.

ASPHODELOIDES, as-fod-el-*oy*-dez, asphodel-like.
SETIFOLIUM, set-if-*o*-le-um, bristle-leaved.

**Yucca,** *yuk*-ka; modification of an aboriginal name applied to another plant. Flowering and foliage shrubs.

ALOIFOLIA, al-o-if-*ol*-e-a, Aloe-like foliage.
FILAMENTOSA, fil-a-men-*to*-za, thready, referring to the filaments on the leaf-margins.
GLAUCA, *glaw*-ka, sea-green—the foliage.
GLORIOSA, glor-i-*o*-sa, glorious.
RECURVIFOLIA, re-kur-ve-*fo*-le-a, recurved leaves.
WHIPPLEI, *whip*-ple-i, after Whipple.

119

**Zaluzianskia,** zal-u-ze-*an*-ske-a; after Adam Zaluziansky, a physician and botanist, of Prague. Half-hardy annuals.
CAPENSIS, ka-*pen*-sis, of the Cape of Good Hope.
SELAGINOIDES, sel-ag-in-*oy*-dez, Selaginella-like.

**Zanthoxylum,** zanth-*oks*-il-um; from Gr. *xanthos,* yellow, and *xulon,* wood, the latter being yellow in some species. Shrubs. Also spelt Xanthoxylum.
AILANTHOIDES, a-lanth-*oy*-dez, ailanthus-like.
BUNGEI, bunj-*e*-i, after Bunge, a Russian botanist.
PIPERITUM, pi-per-*e*-tum, peppery, the ground seeds used as pepper in Japan.

**Zauschneria,** zawsch-*neer*-e-a; after M. Zuaschner, a botanist. Half-hardy shrubby perennial.
CALIFORNICA, kal-if-*or*-nik-a, of California.

**Zea,** *ze*-a; old Gr. name, possibly from Gr. *zea,* a kind of corn. Ornamental foliage, fodder and corn grasses,
MAYS, mayz, old name, probably native. The Indian Corn or Maize.
M. GRACILLIMA, gra-*sil*-lim-a, graceful.
M. VARIEGATA, var-e-eg-*a*-ta, variegated-leaved.

**Zebrina,** ze-*bry*-na; from Zebra, the leaves striped like a zebra. Greenhouse trailer.
PENDULA, *pen*-du-la, drooping.

**Zelkova,** zel-*ko*-va; from the vernacular name in the Caucasus. Sometimes written Zelkowa. Shrubs and trees.
ACUMINATA, ak-u-min-*a*-ta, long-pointed—the leaves.
CRENATA, kre-*na*-ta, notched—the leaves.

**Zenobia,** zen-*o*-be-a; after Zenobia, once Empress of Palmyra. Flowering shrubs.
SPECIOSA, spes-e-*o*-sa, showy.
PULVERULENTA, pul-ver-ul-*en*-ta, powdered—the leaves are glaucous.

**Zephyranthes,** zef-er-*an*-thez; from Gr. *zephyr,* the west wind, and *anthos,* a flower. Half-hardy bulbs.
ATAMASCO, at-am-*as*-ko, old name. The Atamasco Lily.
CANDIDA, *kan*-did-a, white.
CARINATA, kar-in-*a*-ta, keeled.
ROSEA, *ro*-ze-a, rose-coloured.
TREATIÆ, tre-at-e-e, after Mrs. Treat.
VERSICOLOR, ver-*sik*-o-lor, variously coloured.

**Zingiber,** *zing*-ib-er; classical name coming from the Sanskrit. Tropical foliage and flowering plants.
OFFICINALE, of-fis-in-*a*-le, of the shops. The roots form the Ginger of commerce.

**Zinnia,** *zin*-ne-a; named after Johann Gottfried Zinn, a German professor of botany. Half-hardy annuals.
ELEGANS, *el*-e-ganz, elegant.
HAAGEANA, ha-ag-e-*a*-na, after Haage, German nurseryman.
PAUCIFLORA, paw-sif-*lo*-ra, few-flowered.
VERTICILLATA, ver-tis-il-*la*-ta, whorled—the leaves.

**Zizania,** ziz-*a*-ne-a; from *zizanion,* an ancient Greek name for a "tare," or weed. Waterside grasses.
AQUATICA, a-*kwat*-ik-a, aquatic.
LATIFOLIA, lat-if-*o*-le-a, broad-leaved.

**Zizyphus,** *ziz*-if-us; from Zizouf, the Arabian name of Z. Lotus. Shrubs.
LOTUS, *lo*-tus, lotus. The African or Jujube Lotus.

**Zygadenus,** zig-a-*de*-nus, from Gr. *xugeo,* to be joined, and *aden,* a gland, alluding to the double glands of the perianth. Herbaceous perennials.
ANGUSTIFOLIUS, an-gus-tif-*o*-le-us, narrow-leaved.
MUSCITOXICUS, mus-ke-*toks*-ik-us, old name, meaning "fly-poison."
NUTTALLII, nut-*tal*-lei, after Nuttall, a botanist, of Massachusetts.

**Zygocactus,** zy-go-*kak*-tus; cactus with zygomorphic flowers, allied to Epiphyllum. Winter-blooming greenhouse cactus from Brazil.
TRUNCATUS, trun-*ka*-tus, ending bluntly—the stem segments.

**Zygocolax,** zy-go-*ko*-lax; compound of *Zygopetalum* and *Colax.* Warm-house hybrid orchids.
AMESIANA, am-es-e-*an*-a, after F. L. Ames.
VEITCHII, *veech*-ei, after Messrs. Veitch, nurserymen, late of Chelsea.

**Zygopetalum,** zy-go-*pet*-a-lum; from Gr. *zygon,* a yoke, and *petalon,* a petal, in reference to the segments of the perianth being joined at their bases. Warm-house orchids.
CRINITUM, *kryn*-it-um, hairy.
INTERMEDIUM, in-ter-*me*-de-um, intermediate.
MACKAYI, mak-*kay*-i, after Mackay.
MAXILLARE, max-il-*la*-re, jaw-shaped.
SEDENII, sed-*e*-nei, after John Seden, a noted hybridist.